Technological Innovation

Related title from Palgrave Macmillan

KNOWLEDGE CREATION PROCESSES: Theory and Empirical Evidence from Knowledge-Intensive Firms (by Gregorio Martín de Castro *et al.*)

Technological Innovation

An Intellectual Capital-Based View

Gregorio Martín de Castro
Associate Professor of Business Administration, Universidad Complutense de Madrid, Spain

Miriam Delgado Verde
Assistant Professor of Business Administration, Universidad Complutense de Madrid, Spain

Pedro López Sáez
Associate Professor of Business Administration, Universidad Complutense de Madrid, Spain

and

José Emilio Navas López
Professor of Business Administration, Universidad Complutense de Madrid, Spain

First published 2010 by
PALGRAVE MACMILLAN

Palgrave Macmillan in the UK is an imprint of Macmillan Publishers Limited, registered in England, company number 785998, of Houndmills, Basingstoke, Hampshire RG21 6XS.

Palgrave Macmillan in the US is a division of St Martin's Press LLC, 175 Fifth Avenue, New York, NY 10010.

Palgrave Macmillan is the global academic imprint of the above companies and has companies and representatives throughout the world.

Palgrave® and Macmillan® are registered trademarks in the United States, the United Kingdom, Europe and other countries.

ISBN 978-0-230-23021-7

This book is printed on paper suitable for recycling and made from fully managed and sustained forest sources. Logging, pulping and manufacturing processes are expected to conform to the environmental regulations of the country of origin.

A catalogue record for this book is available from the British Library.

Library of Congress Cataloging-in-Publication Data
Technological innovation : an intellectual capital-based view / Gregorio Martín de Castro ... [et al.].
       p. cm.
    1. Technological innovations—Management.   2. Knowledge management.
I. Martín de Castro, Gregorio, 1972–
HD45.T3944 2010
658.5'14—dc22

                                                                  2010010817

10   9   8   7   6   5   4   3   2   1
19   18   17   16   15   14   13   12   11   10

Contents

List of Figures

List of Tables

Acknowledgements

The authors wish to acknowledge all the people and institutions who have believed in our project and have made it possible in any way. First, to Professor Jeremy Howells, for his ongoing support during our research fellowship at the Manchester Institute of Innovation Research, University of Manchester (UK). To the Rafael del Pino Foundation (Spain), and its director, Amadeo Petitbò Juan, for trusting in the work and funding the research. We also wish to thank to Jorge Cruz González, researcher and PhD candidate at the Business Administration Department, Universidad Complutense (Spain), for his valuable assistance in data-gathering process, and for his commitment and efforts. Finally, the authors would like to express their gratitude to all Spanish managers and organizations who have devoted their time and effort to this research, turning it into a reality. To all of them, thank you very much.

Foreword

In recent years knowledge and intellectual capital associated with its subsequent development and commercialization have been recognized as major elements in the economic and business world. However, this recognition and acknowledgement have only gradually been reflected in research, or more particularly in new approaches and models surrounding these developments. This book aims to remedy such a situation by focusing on the intangible resources and capabilities firms use when competing in the newly emerging global economy. As such, firms increasingly recognize that growth and development are dependent on effectively harnessing knowledge and related intangible resources to compete against other companies. The book charts how firms can manage this process as well as providing a new conceptual framework for our understanding of this key strategic and managerial process.

The book explores this notion more specifically by advancing a new theoretical approach, the Intellectual Capital-Based View (ICV) of the firm, which seeks to provide a more grounded and testable model of strategic management and development than the existing Resource-Based View (RBV) of the firm. This new approach identifies different elements of knowledge within a firm and relates this to its intellectual capital base. These knowledge stocks or blocks highlight different elements of the intellectual capital base of a company. However, the work seeks not only to present the model of the ICV approach but also to present ways in which it can be measured empirically, something which has been lacking in previous studies. This has been particularly true with respect to the RBV approach to competitive advantage where intangible resources and capabilities of the firm, although recognized as increasingly important, remained generalized and vague.

The study then goes on to relate this ICV approach to the innovation dynamics of the firm and how knowledge and intellectual capital needs to be effectively harnessed to meet the needs of ever demanding consumers and the wider challenges of the increasingly complex and uncertain task environment of the firm. As part of this investigation the book analyses the empirical link between the components of intellectual capital of the firm and its innovative performance centred on its product and process innovation profiles.

By successfully unpacking the notion of intellectual capital and the knowledge assets of the firm the book provides a major contribution to our understanding of the new competitive dynamics and frontiers of the firm in the twenty-first century. The book therefore deserves to be widely read, explored and tested by both managers and theoreticians who are seeking to move beyond the traditional and largely static and unquantifiable strategic models of the past.

JEREMY HOWELLS
Professor in the
Manchester Institute of Innovation Research,
University of Manchester

Introduction

An intense process of social and economic change taking place over recent years has led to a new environment. The globalization of economic activity, the advances related to technological revolution, the increasing relevance of service industries in the economy, accelerated product cycles, changing customer preferences and needs and so on are some of the main phenomena to take into account in the business landscape. As a consequence, the evolution of economic activity is directed to the so-called 'knowledge society' (Grant, 1996; Dean and Kretschmer, 2007), which is directly related to the creation, use and exchange of knowledge.

In this sense, knowledge has an increasingly important role as a strategic resource in business competition (Grant, 1996; Teece, 1998). However, the organizational capability of a firm to build and reinforce its competitive position is a key issue that is necessary to consider when analysing that term.

Therefore, the new competitive dynamic (Johnson *et al.*, 2002; Leitner, 2005) generated by the new socioeconomic environment leads to paying attention to intangible resources and capabilities when firms are facing competitors. In short, competition based on intangible resources is the result of the incorporation of knowledge into the different organizational, productive and management activities, as well as to the range of products and services offered.

From this situation, new theoretical approaches have appeared within the academic landscape, attempting to explain the nature and sources of competitive advantage (one of the main research topics in the field of Business Management) from an internal point of view.

In this vein, over more than two decades, the explanation of the creation and sustaining of competitive advantage has been tied to the possession and/or control of firms' endogenous factors, taking into account in particular the intangible ones (Itami and Roehl, 1987; Hall, 1993). For this reason, the resource-based view (RBV) and the intellectual capital-based view (ICV) (Reed *et al.*, 2006) will be considered as the main theoretical approaches, since they are some of the most widely accepted theoretical perspectives in the strategic management field (Powell, 2001), emphasizing the role of firms' internal factors in obtaining and maintaining positions of competitive advantage.

Specifically, the main focus of this book will be intangible factors, based on information and knowledge, since they are more likely to contribute to attaining and sustaining a competitive advantage. In this sense, we pay attention to an emerging factor of production – intellectual capital – which supplements/replaces land, labour and capital (Dean and Kretschmer, 2007).

Subsequently, we focus on the intellectual capital-based view of the firm because the actual level of empirical support for the resource-based view remains uncertain (Newbert, 2007). Some of the RBV's concerns refer to 'how ... we conceptualize and then measure a concept that is based on some firm-specific interaction of resources, which themselves are intangible, and therefore, unobservable?' (Reed *et al.*, 2006: 868). These authors assert that the RBV's lack of specificity make it difficult to design and test empirically.

Hence, even though we start from the RBV, we consider ICV in order to solve these concerns, as it provides a higher potential for empirical testing, representing specific aspects more narrowly linked to a firm's competitive advantage.

The ICV approach arose from practitioners in the 1990s and early 2000s (Brooking, 1996; Edvinsson and Malone, 1997; CIC, 2003) and distinguishes between different blocks of intellectual capital or types of organizational knowledge stocks. That is, those intangible resources and capabilities – which are knowledge in essence – are embodied in different forms within the firm, giving rise to the different components of intellectual capital. In this vein, they can be analysed as human capital, structural capital and relational capital, in accordance with most of the studies. In addition, each element refers to a different type of knowledge: employees' knowledge, organizational knowledge and inter-organizational knowledge.

Nevertheless, in spite of the great number of studies that analyse intellectual capital and its elements, nowadays 'a phenomenon as complex as intellectual capital, requires comprehensive theoretical and empirical development' (Cabrita and Bontis, 2008: 214); and this is where our research aims to contribute. It is because of a shortage of empirical studies that analyse the dimensions of each of the elements of intellectual capital and contribute appropriate measurement scales.

Therefore, assuming that the models of intellectual capital have become highly relevant, because they not only allow us to understand the nature of these assets, but also to carry out their measurement, our study tries to advance the consideration of dimensions within each element of intellectual capital as well as the measurement of such dimensions.

On the other hand, several authors highlight the relevant role played by innovation activities (Schumpeter, 1942; Tushman and Nadler, 1986; Van de Ven, 1986; Alegre *et al.*, 2005; Salman and Saives, 2005; Galende, 2006). Thus firms should create new products, services and processes to be able to compete in the present environment, and they should take innovation as a business model in order to master that environment.

In this sense, in the current competitive and dynamic environment, technological innovation is becoming increasingly a key aspect for business competition. In fact, many studies have shown that technological innovation can entail positive impacts, increasing the competitiveness of a firm (Yam *et al.*, 2004). For this reason, studies focused on improving innovative firm processes are of major interest.

It is therefore necessary to accumulate a large amount of knowledge in order to develop different types of innovation. Thus the growing importance of knowledge as a key productive factor and central issue for innovation can be explained by the continuous accumulation of technical knowledge over time and the utilization of communication technologies that quickly spread knowledge throughout the world. In this way, knowledge and innovation are dominant resources in the contemporary knowledge-based economy (Tseng and Goo, 2005), and are a key argument for carrying out empirical research such as ours.

In addition, in the professional field, managers need tools to evaluate their management and the outcomes of innovation activities. Thus, because of consumers' complex and changing demands, which have led to an increasing range of products and services, it is fundamental to emphasize the relevance of the innovation process in the present business landscape, so knowledge as well as its applications are key elements in achieving and maintaining business success.

Focusing on the relationship between intellectual capital and innovation, there are many studies arguing that innovations come from intellectual capital (Edvinsson and Sullivan, 1996; Bontis, 1998; Nahapiet and Ghoshal, 1998; Tsai and Ghoshal, 1998; Sullivan, 2001; Hermans and Kauranen, 2005; Subramaniam and Youndt, 2005; Swart, 2006, among others). Nevertheless, there is a lack of empirical studies that analyse in depth the relationship between the different elements of intellectual capital and different types of innovation.

In this sense, this book attempts to provide an empirical contribution regarding those relationships, following the commentaries carried out by authors such as Sullivan (2001), Darroch and McNaughton (2002) or Tödtling *et al.* (2009), who point out the shortage of empirical research on this subject. In addition, an empirical study with these characteristics will provide evidence about what managerial actions should be taken by firms when dealing with their intellectual capital, identifying the knowledge and processes with a higher potential for different successful innovations. This is an important issue because, according to McEvily *et al.* (2004), different types of innovation will require the use and management of different resources within the firm.

Based on the comments above, this book tries: (i) to identify and classify the different elements of intellectual capital as well as the dimensions that make up these elements; and (ii) to throw some more light on the relationships between intellectual capital and technological innovation, analysing the influence of each of the components of intellectual capital (human capital, structural capital and relational capital) on two main types of technological innovation (product and process innovation).

To achieve these objectives, the book has been structured as follows: In Chapter 1, the theoretical background is developed, considering the theories or approaches that sustain the importance of intangible factors in achieving sustained competitive advantage in a firm – namely the resource-based view, the knowledge-based view, and the intellectual-capital-based view. In this sense, we attempt to analyse the significant aspects of the new socioeconomic environment that affect business competition and characterize it, stressing intangible resources and capabilities owned by the firm, which that are knowledge in essence, and establishing the research bases. In addition, the important role that knowledge plays in a company is emphasized, and the arguments considered for undertaking our research efforts are presented, trying to show the mechanisms that allow us to understand and frame intangible and internal factors.

In Chapter 2, an analysis of previous research on intellectual capital is carried out, considering the main works that examine it. In this vein, the components of intellectual capital and the dimensions into which each of the components can be divided are studied, emphasizing some general agreement with respect to the main kinds of intellectual capital and trying to homogenize all ideas considered within such studies to achieve a more complete and precise model. A model of intellectual capital emerges from this review effort, considering the blocks of intellectual capital (human capital, structural capital and relational capital) as the dimensions and providing the most consistent classification according to the theoretical review, since the works considered refer to intangible factors that are not included within the intellectual capital literature.

In Chapter 3, several studies from the field of innovation, particularly focused on technological innovation, are examined theoretically. The aim is to show the paradigm held by the academic literature, analysing the concept as well as its main typologies. On the one hand, based on the literature, the importance of an ongoing adaptation to environment changes is noted, emphasizing subsequent innovations developed by a firm in order to achieve a competitive advantage. On the other hand, from identified different classification criteria, the choice of two types of innovation are justified to carry out the empirical study, focusing on product and process innovation, which are the typologies most accepted by the academic literature. In addition, we try to show its importance, taking into account some of the main works belonging to the field of innovation.

In Chapter 4, the hypotheses for the empirical research are developed. Thus, from those empirical studies that take intellectual capital (or intangible factors) as a source of technological innovation, hypotheses about the different relationships between each element of intellectual capital and the two types of technological innovation are raised, providing several arguments to support the six hypotheses. In this sense, in this book it is expected that each component of intellectual capital enhances technological innovation, and thus proposing positive relationships among them.

In Chapter 5, our methodology is described, emphasizing previous empirical studies focused on this topic, as well as those contributions regarding methodological indications to follow, such as that related to choosing industries according to their characteristics, designing questionnaires, and issues related to the sample of the study. With respect to companies considered, we focus on high- and medium-high-technology Spanish manufacturing firms with fifty or more employees, because such companies may have somewhat formalized innovation systems and a complex intellectual capital structure. In this way, we provide details about those industries that were included in the study as well as explaining why we pay attention to these sectors. Regarding our work, an 'ad hoc' questionnaire was used as the basic tool for collecting information, allowing us to obtain data directly from the day-to-day business reality that firms face. A series of dimensions and variables were designed, which were used as the basis for the elaboration of a questionnaire with Likert 1–7 scales. These were used for gathering primary data. Finally, it is interesting to stress that the fieldwork finished in May 2009, having lasted approximately two months, and the questionnaires were aimed at top management.

Chapter 6 describes the statistical techniques used in order to test the hypotheses previously raised. The main results of the empirical investigation are shown, once the information had been obtained and processed statistically. On the one hand, statistics were used for validating the proposed measurement model, and on the other, for verifying or rejecting the causal relationships stated in the hypothesis. In this sense, an exploratory and confirmatory factor analysis was carried out to determine the components of intellectual capital as well as technological innovation. A multiple regression model was also designed, to test the effects of each kind of intellectual capital on the two different types of innovation.

Finally, in Chapter 7, the most relevant conclusions drawn from the theoretical review and previous empirical results, and future research directions tied to this study, are shown. In this way, both theoretical and empirical findings and contributions are discussed, and recommendations for managerial practice extracted from the results.

1
Theoretical Framework

In this first chapter we shall try to provide a coherent framework in order to understand the phenomena studied. Basically we shall use the two main theoretical frameworks most widely used by academics in order to explain technological advantage: (i) the knowledge-based view of the firm; and (ii) the resource-based view of the firm. We shall also include a more pragmatic view and middle-range theory: the intellectual-capital-based view of the firm.

Knowledge, strategy and the firm

Zack (1999) argued that economic organizations are recognizing knowledge as a very valuable and strategic resource. Therefore, realizing that their competitiveness largely depends on it, they know it is necessary to make decisions that allow them to manage their intellectual resources and capabilities. Teece (1998) agreed with these sentiments, since he considered that economic prosperity is based on knowledge and its useful application. He considered that structural changes experienced by the economies of developed countries have determined what is strategic, and have underlined the importance of knowledge and its management.

Since the mid-1990s, the growing interest among firms regarding learning and knowledge has entailed an increase in several streams of analysis, based on managerial practice as well as in the academic domain, that have been devoted to understanding this phenomenon. In general, these could be grouped into: (i) the analysis of knowledge stocks, intangible resources and capabilities, or the intellectual capital of the organizations; (ii) the flows that create and transfer knowledge within and between organizations, or simply organizational learning as a dynamic reality; and (iii) the management efforts and decisions

1

devoted to foster and control these flows of explicit knowledge and information inside the firm.

All this interest about knowledge as a key competitive factor has led to the development of an entirely new theory of the firm. This theory (Kogut and Zander, 1992; Nonaka and Takeuchi, 1995; Grant, 1996; Spender, 1996) claims that organizations have particular capabilities for knowledge creation and sharing which confer a comparative advantage, in terms of economic efficiency, in comparison with other institutional forms as markets (Nahapiet and Ghoshal, 1998). Thus, the knowledge-based view of the firm explains the existence, growth, evolution, boundaries and internal organization of the firm, taking as the key point its available knowledge.

For strategy theory, this perspective represents a shift in the focus of analysis from the historically dominant focus based on value appropriation to one based on value creation through knowledge and learning. Conventional theories of the firm have concentrated on opportunism to a great extent. This phenomenon has been applied in explaining the existence of the firm, arguing that organizations are mechanisms to reduce the costs associated with opportunism that are found in the market. In addition, the costs related to opportunism that arise inside the firm are the main limit to its growth, and define its boundaries for cost reduction.

In the knowledge based view of the firm, while the basic dichotomy between self-interest and a yearning for participation in a community is still alive, the focus on collaboration is employed as the main support for claiming organizational superiority over markets, because organizations solve this fundamental problem of ongoing collaboration between individuals (Kogut and Zander, 1996).

Kogut and Zander (1996) argued that firm knowledge entails a higher economic value than market transactions when the shared identity of individuals within the organization creates a social knowledge that sustains and fosters co-ordination and communication. When individuals share an organizational identity they internalize behavioural rules, and co-ordination and communication becomes easier among individuals and groups with specialized abilities.

Organizations provide a feeling of social community to individuals, and through this community and shared identity, collective co-ordination and learning are achieved. Therefore, the reason for a firm's existence does not rest on smaller costs compared to those of the market, but in the fact that it provides an appropriate context for creating social and/or collective knowledge that is impossible to obtain in the market, where this shared feeling of identity does not exist. Thus, specialized knowledge that is not available in the market in the normal way because it is socially constructed is the only feature that creates firm value in the knowledge economy.

The knowledge-based view of the firm is an economic theory that has been developed since the mid-1990s, and it appears that it will be one of the key issues in the fields of strategy and organization during the years to come. The theoretical body of the knowledge-based theory of the firm (Conner, 1991; Kogut and Zander, 1992, 1996; Conner and Prahalad, 1996; Grant, 1996; Spender, 1996; Foss, 1997; and more recent authors) comes from several research streams. A brief review of the theoretical roots of the new theory is provided in the following subsections. The key assumptions and main elements of the theory are highlighted, pointing out how these elements relate each other, and how they flow from the basic assumptions of other research streams.

The knowledge-based theory of the firm is the result of several modifications to traditional economics conceptions. Thus, in a general manner, it can be said that it comes from the convergence of the following research streams: (i) the resource-based view; (ii) epistemology; (iii) the evolutionary perspective; (iv) technological innovation; (v) organizational learning; and (vi) the social systems theory.

The resource-based view

The resource-based view and the knowledge-based view of the firm share the primary goal of exploring firms' internal dynamics; to dismantle the 'black box' and discover its main components and interactions (Spender, 1996). Thus both heterogeneity among firms along knowledge bases, and the sustainability of these differences because of the specific and historically dependent process of learning are at the core of knowledge-based thinking (Conner, 1991; Dossi and Teece, 1993; Foss, 1994, 1996; Cohendet *et al.*, 2000; Dossi and Marengo, 2000).

The knowledge-based view of the firm, as an extension of the resource-based view (Conner, 1991), accepts the different criteria or requirements developed by the latter as an intermediate step in analysing ways of gaining and sustaining a competitive advantage among firms. But the former also applies these criteria partially to the competition between markets and firms as different ways of organizing economic activity.

Because of its outstanding relevance for the subject of this book, the next section of this chapter is dedicated to the resource-based view.

Path dependence and evolutionary perspectives

One of the main elements that the knowledge-based view of the firm takes from the resource-based view is firm heterogeneity. This heterogeneity can be analysed through the epistemological and ontological nature of the knowledge base of each firm.

The other borrowed element is sustainable differences over time, which makes it necessary to introduce dynamic mechanisms to the developing theory. At this point, the contributions from path dependence theory and evolutionary approaches become critical. These perspectives are often so close to the knowledge-based view of the firm that the authors of the competencies approach, organizational capabilities and dynamic capabilities (Nelson and Winter, 1982; Dossi and Teece, 1993; Foss, 1994, 1997; Teece *et al.*, 1997; Dossi and Marengo, 2000) are frequently taken for parallel theoretical developments.

As Foss (1994) stated, a theory of path dependence shows how historical facts lock-in a social path for future development. This argument can easily be applied to firms. The knowledge base of a firm at a given moment is a consequence of a set of historical events, and the result of learning experiences from the past (Barney, 1991). In turn, this knowledge determines a 'learning path' for the future. In other words, where a firm can go is a function of where it has been (Dossi and Teece, 1993).

Markets also constitute a selection mechanism that chooses among different firm behaviours though its outcomes (Foss, 1994). The evolutionary perspective is focused on replication, variation and selection mechanisms that are applied to the knowledge available inside a firm. These mechanisms also rule market dynamics and firm survival through learning and adaptive capabilities (Nelson and Winter, 1982; Kogut and Zander, 1992; Teece *et al.*, 1997).

Technological innovation and the management of technology

The fields of technological innovation, technology management and new product development (Teece, 1986; Henderson and Cockburn, 1994; Patel and Pavitt, 1995; Cowan and Foray, 1997; Saviotti, 1998; DeCarolis and Deeds, 1999; DeCarolis, 2003) also maintain important links to the previously mentioned research streams, and with the knowledge-based theory of the firm.

The process of technological innovation is, by its very nature, evolutionary and historically dependent. It represents the breaking of the classical assumption of a constant and public technological function. This shift to 'Schumpeterian shocks' is at the heart of firm heterogeneity (Dossi and Teece, 1993), and it confers learning roles to economic agents (Boerner *et al.*, 2001).

Another key idea that comes from the innovation literature is that of the complex nature of technology. This complexity means that tacit, collective and highly firm-specific knowledge creates a need for the development of internal mechanisms to manage a range of factors and interactions (Boerner *et al.*, 2001), and this ability is, by itself, a complex kind of procedural knowledge.

Decisions about technology management will be influenced strongly by the currently available knowledge base, which constrains and guides these decisions in a lock-in fashion (Dossi and Teece, 1993; Kogut and Zander, 1993, 1995). In addition, new product development is also related to new knowledge creation and the exploitation of already available knowledge, in a search for value creation (Kogut and Zander, 1995; Nonaka and Takeuchi, 1995).

Because of the relevance that technological innovation holds for this book, Chapter 3 will address its main concepts and components.

Organizational learning and the knowledge-based theory of the firm

Contributions from the organizational learning arena (Lounamaa and March, 1987; Argyris, 1991; Huber, 1991; Dodgson, 1993; Levinthal and March, 1993; Crossan *et al.*, 1999) are focused on how available knowledge is accumulated through time and space. These authors devote themselves mainly to analysing the different stages that can be found in this process, the different factors that foster or constrain it, and to the proposition of learning categories.

If we adopt the perspective of learning based on behaviours that Huber (1991) proposed, we can say that an entity learns when, through knowledge processing, the extent of its potential behaviours changes. The connection between organizational learning and evolutionary perspective is clear. Learning is a dynamic process that connects past, present and future knowledge, with the platform and barriers that path dependence builds for future learning.

Clearly, learning also includes the relationship between available knowledge (resources) and potential behaviour (performance) that is explored in the resource-based view of competitive advantage, though, obviously, in this case the learning entity has to be a firm.

Thus the concept of organizational learning is needed in order to understand the connections of this research stream with the rest of the sources of the knowledge-based view of the firm. Organizational learning exists when any of the units that are part of the organization acquire knowledge recognized as being potentially useful for the firm, and such learning is wider when more of the organizational components have access to potentially useful knowledge (Huber, 1991). This means that the process of organizational learning can take place at any ontological level and involving any kind of knowledge (tacit or explicit), and that its extent depends on the number of levels and kinds of knowledge involved.

Social systems theory and the knowledge-based theory of the firm

The final source of the knowledge-based view of the firm is the social systems theory. This can include network perspectives as well as the approaches focused on the social and cultural identity of the firm (Kogut and Zander, 1996; Brown and Duguid, 1998; Nahapiet and Ghoshal, 1998; Lipparini and Fratocchi, 1999; Kogut, 2000).

Spender (1996) claimed the need for a holistic point of view in order to study organizations, and recommended the use of systems theory because the firm is usually involved socially, technically, politically and socially.

The distinction between 'component knowledge' (needed by the subsystems) and 'architectural knowledge' (which is necessary to connect the different subsystems) becomes critical (Grant, 1997), because once the different agents involved in a certain architecture are identified and selected, other attributes are needed. These system attributes include competence combination, functional co-ordination and systemic integration (Lipparini and Fratocchi, 1999).

These systemic properties or architectural knowledge are specifically tied to obtaining synergies that come from the use of co-ordination. The knowledge-based theory of the firm includes the synergistic potential of several individuals working together (Brown and Duguid, 1998). This potential synergy resulting from the combination of complementary abilities or kinds of knowledge can only be released if people learn to work as a team, and if the routines and the know-how for appropriate co-ordination are present, which denotes the dynamic earnings of team value creation (Madhok, 1996).

The resource-based view of competitive advantage

Despite the study on the nature and sources of firm success being one of the most outstanding topics in the field of business management, since the 1970s we have seen an important shift of focus in order to carry out this study. Until the final years of the 1980s, the dominant paradigm was the focus on Structure–Behaviour–Results, according to the fundamental arguments of industrial organization and of Michael Porter (1980, 1985). Following this scheme, the origin of sustained competitive advantage of organizations was located in exogenous factors that cannot be controlled directly by any firm.

In the decade of the 1990s powerful new focal points arose, claiming the relevance of internal factors, and arguing that the true source of a firm's success lies in several idiosyncratic and heterogeneous assets that constitute the particular characteristic of the organization. These contributions appeared initially under the umbrella of the resource-based view (Wernefelt, 1984; Barney, 1991; Grant, 1991; Peteraf, 1993), and later, were extended to become the knowledge-based view of the firm (Nonaka and Takeuchi, 1995; Grant, 1996; Kogut and Zander, 1996; Spender, 1996).

While the results of empirical research still show an element of debate, there is general agreement that the differences in business performance observed among firms in a same industry are wider and more persistent over time than those that can be found among firms operating in different industrial settings (Rumelt, 1991; McGahan and Porter, 1997, 1998, 2002). This can be a first clue about the relevance of internal firm factors, in particular those that are based on knowledge, and its role as the main determinants of differences in business performance.

These arguments have been developed since the 1980s within the framework of the resource-based view (Wernerfelt, 1984; Barney, 1986, 1991, 2001; Dierickx and Cool, 1989; Grant, 1991; Amit and Schoemaker, 1993; Peteraf, 1993) and the knowledge-based view (Nonaka, 1991; Kogut and Zander, 1992; Hedlund, 1994; Nonaka and Takeuchi, 1995; Zander and Kogut, 1995; Grant, 1996; Spender, 1996; Sánchez, 2001), and have pointed out that organizational factors with intangible properties, or simply knowledge-based (Itami and Roehl, 1987), can be the main source of sustained competitive advantage. In this sense, we must highlight that these frameworks are not exclusively internal, because some manifestations of organizational knowledge (Kogut and Zander, 1992; Drucker, 1995; Grant, 1996; Spender and Grant, 1996) are considered to be key elements in order to face the changing conditions of the external environment.

Obtaining competitive advantage

When determining the nature and sources of competitive advantage, it is necessary to distinguish two phases: the first relates to how a firm achieves an advantage over its competitors; and the second addresses how a company can sustain or defend its position over time, while obtaining returns over the industry mean.

With regard to the analysis of the sources from which to obtain competitive advantage, the previous section has noted that unique and valuable knowledge allows the firm to outperform the market as well as its competitors. Thus the knowledge base of the firm and its learning capabilities, in particular when they are better and quicker than those of the competition, can become the most valuable resource of a company. The idiosyncratic and heterogeneous character of knowledge, as well as superior means of creating and leveraging it, constitute one of the basic routes to obtaining competitive advantage.

Following the arguments of Michael Zack (1999), it can be said that the strategic context of an organization helps to identify the knowledge-based initiatives that support its strategic purpose or mission, reinforcing its competitive position and creating value for shareholders.

In an intuitive way, it makes sense to think that the firms with a better understanding of their customers, products, technologies, markets and organizational relations will perform excellently. Therefore, knowledge should be considered strategically as the most important resource, and the capability to acquire, integrate, store, share and apply it, will be the most valuable organizational means of gaining and sustaining competitive advantage (Grant, 1996).

Knowledge-based resources rise from the accumulated expertise and know-how possessed by individuals; firms add a physical, technical and social structure in order to use this knowledge to shape organizational capabilities, and employing it to provide goods and services. The way in which a firm configures, deploys and redeploys its knowledge-based resources, competencies and capabilities will be a critical determinant of economic performance and the firm's commercial success. Therefore, in the contemporary knowledge economy, the competitive advantage of the firm does not come so much from market power, but rather from knowledge-based assets and the way in which they are deployed (Teece, 1998). This iterative and ongoing process of deployment, combination, redevelopment and recombination represents the dynamic considerations embedded in the research streams of knowledge management and organizational learning.

Nevertheless, once this kind of advantage has been achieved, it is necessary to analyse its chances for maintenance over time, which would constitute the second phase in the strategic analysis of competitive advantage according to the resource-based view.

Sustaining competitive advantage

The main reasons why the specialised literature has mentioned sustaining resource-based competitive advantage, the effects of 'mass efficiencies' and 'diseconomies in time compression' needs to be highlighted (Dierickx and

Cool, 1989; Grant, 1991). These particular effects are tied to the nature of knowledge as an essential production factor. A synthesis of both phenomena in simpler terms can be found in the work of Zack (1999): 'the more things a firm knows, the more things it can learn'.

To be able to observe the effects of path dependence, or simply the consequences of the historical timeline in resource accumulation, it is necessary to adopt a dynamic perspective for analysis. Thus we must move briefly towards an organizational learning and process-based perspective, taking flow variables for study, and leaving aside the stock variables that are closely tied to the analysis of intellectual capital. In this way, our focus moves to the creation and development of a firm's knowledge and capabilities.

An analysis focused on learning allows an understanding of the means by which a firm can reach a certain knowledge stock. Therefore, we can see that the link between both perspectives lies in the fact that the knowledge stocks of the firm represent the inputs as well as the outcomes of the organizational learning process.

This is probably the core of the resource-based approach. Intangible resources and organizational capabilities have a common essence: both are knowledge-based, or even are themselves a form of knowledge. If the resources and capabilities are available to the firm are responsible for situations of competitive advantage or disadvantage and, at the same time, they are a representation of the firm's knowledge stocks, domains or endowments, then these knowledge or intellectual capital stocks develop and constrain firm possibilities for reaching a competitive advantage.

In practice, the effective maintenance of business success depends on, among other aspects, barriers to imitation, transfer and/or substitution of the competitive advantage (Barney, 1991). In this sense, capabilities built upon 'socially complex' knowledge are usually very difficult to replicate, transfer or substitute. This fact is based fundamentally on the well-known concept of 'causal ambiguity' (Lippman and Rumelt, 1982; Reed and DeFillippi, 1990), which refers to the impossibility that competitors face in determining accurately the relations that appear between owning or controlling a certain group of organizational factors and the business performance that the firm achieves.

These processes can also be hindered because, in most cases, these types of idiosyncratic resources and capabilities are not developed in an isolated way, but act in a co-ordinated manner, showing a high degree of complementary relations and co-specialization. Thus, organizational capabilities configure networks of lower or higher complexity through a whole set of a firm's capabilities, and these networks can appear quite obscure to competitors or potential imitators (Amit and Schoemaker, 1993). This, evidently, also limits the transfer possibilities for organizational capabilities (Vicente-Lorente, 2001).

Similarly, returning to the work of Zack (1999), it can be argued that knowledge-based resources can be unique and very difficult to imitate, in particular when their embodied knowledge is (a) context specific; (b) it presents a high degree of tacitness; (c) it is embedded in complex organizational routines; and/or (d) it is developed directly from practical experience.

Contrary to the case of many traditional resources, knowledge may not be acquired easily through markets. This can make being involved in similar experiences necessary for competitors interested in acquiring similar knowledge, as the only way to replicate specific and idiosyncratic knowledge. However, acquiring knowledge by means of practical experience takes time, and competitors face important limitations on the possibility of learning enhancement through increases in their expenses or investment levels. This opens a window for knowledge-based competitive advantage sustainability over time, but also makes it necessary to include in the analysis all these dynamic or longitudinal processes.

A comprehensive synthesis of the resource-based view

At present, the resource-based view (RBV) of the firm is perhaps one of the most influential frameworks in the strategic management landscape. Nevertheless, several weaknesses can be found in this framework. One of them is about the empirical basic criteria that resources have to fulfil in order to be considered as 'strategic resources'. This point is one of the foundations of Priem and Butler's (2001) 'tautology criticism'. In his work, Barney (2001) made reference to the parameterization of value and rareness. As Barney (2001: 44) commented: 'in most empirical and theoretical work on rarity since the 1991 article, researchers have either implicitly focused on the competitive implications of valuable and unique resources or have been rather imprecise in specifying how rare a resource must be among competing firms to still generate competitive advantages'.

Inimitability has received enough attention in the specialized literature and can be considered to be well parameterized. The origins of inimitability can be tracked to 'unique historical conditions' (Dierickx and Cool, 1989), 'causal ambiguity' (Reed and DeFillippi, 1990) and 'social complexity' (Barney, 1991). Thus, inimitability becomes the main source of sustained competitive advantage (Barney, 2001).

Martín de Castro *et al.* (2009) proposed a set of criteria, based on Dierickx and Cool (1989), Reed and DeFillippi (1990), Barney (1991) and Grant (1991), that try to overcome the mentioned operative barriers to make a direct test for the RBV. These criteria – value, heterogeneity, and complexity – constitute a stage previous to the well-known causal analysis regarding the relationship between resources and sustained superior rents.

Value

Value and heterogeneity are the two basic requirements needed to obtain a competitive advantage. The value of a certain resource is determined by its ability to be used in conceiving or implementing strategies that improve business efficacy and efficiency. Value is determined by the prevailing environmental conditions, and depends on spatial and temporal dimensions (Priem and Butler, 2001). In this sense, the value criterion is influenced by the particular business activity conditions at a given moment in time.

This variable is well explained by Miller and Shamsie (1996). Their paper identified different key organizational factors in Hollywood Cinema Studies – as company reputation or star actors – and studies their evolution over a long period of time. In this longitudinal study, they tested empirically how environmental conditions – exogenous factors – change the value of organizational factors – endogenous factors.

As can be easily deduced, the value criterion is highly influenced by environmental conditions that change over time and among industries. This theoretical proposal is similar to the 'industry strategic factors' presented in the early 1990s by Amit and Schoemaker (1993). To summarize, the value of a firm resource is determined to a great extent by the environmental – temporal and spatial – conditions in which a firm acts.

Degree of heterogeneity

The second basic criterion is resource scarcity or rareness. According to the critiques of Priem and Butler (2001), this issue is not defined clearly. Following the initial proposal of Barney (1991), it is very difficult to determine the maximum number of competing firms that can possess a particular factor for it to be considered rare. The immediate response is that if only one competing firm possesses a particular resource – whether valuable or not–, then this factor will be rare.

We can assume the argument of rareness, but it is difficult to assess. Nevertheless, this problem can be solved by converting a dichotomy issue (rare/not rare) into a matter of degree. In this way, we can understand rareness in terms of a heterogeneous resource endowment among competing firms, and this lies precisely at the heart of the RBV reasoning. Deephouse's (2000) work on the strategic role of corporate reputation can help us with this argument. The author employed simple statistical dispersion measurements for the 'media reputation index' of commercial banks in order to assess the heterogeneity of this factor endowment among firms. Following this idea, the higher the statistical dispersion of a certain factor (reputation, technological knowledge, managerial skills and so on), the greater will be the heterogeneity across firms in relation to that factor endowment. If only a few firms in an industry have a certain organizational factor, then this kind of asset is considered to be heterogeneously available in the industry.

Degree of complexity

Most of the contributions to the RBV framework have analysed the nature and sources of sustained competitive advantage (Roberts and Dowling, 2002), focusing on the inimitability criterion. The parameters for exploring inimitability usually comprise three basic variables: (i) 'unique historical conditions' or 'path dependence'; (ii) 'causal ambiguity'; and (iii) 'social complexity' (Dierickx and Cool, 1989; Reed and DeFillippi, 1990; Barney, 1991). Nevertheless, most of the time these variables appear to be unconnected and without a clear line of reasoning. For this reason, additional work is needed to complete the parameterization of inimitability.

'Causal ambiguity', namely the impossibility of determining clearly the contribution of a certain resource/capability to competitive success, is one of the main sources of inimitability. Following Reed and DeFillippi's (1990) proposal, the origins of causal ambiguity are: (i) number of factors involved; (ii) their relationships; and (iii) the tacitness of these relationships. While this analysis is fairly appropriate, 'causal ambiguity' can also be seen as a result or effect of complexity, or just as a consequence of 'social complexity' – in Barney's words – inherent to a certain organizational factor.

Complexity is the origin of specificity, and it makes a specific resource very difficult to transfer. The better (or even the only) choice for obtaining this kind of firm resources effectively is internal accumulation (Dierickx and Cool, 1989; Hall, 1992), though this involves a long period of time and a path-dependent process. According to these arguments, assessing the degree of complexity of a certain factor should represent an empirical direct test regarding its degree of inimitability and transferability.

Based on the previous ideas as well as on Grant (1991), resource complexity depends on the number of involved factors or elements, as well as on the inter-relationships among them. These two elements of complexity, in addition to the 'tacitness' of the mentioned inter-relationships shape the notion of 'causal ambiguity' presented by Reed and DeFillippi (1990). Thus, as has already been remarked, causal ambiguity is the direct effect of resource complexity.

The intellectual capital-based view of competitive advantage

One of the most powerful streams that appeared in the decade of the 2000s was the perspective of intellectual capital, or the efforts to identify and measure the assets, resources, capabilities or organizational factors with an intangible nature that, in spite of not being taking into account by the traditional accounting systems, can be a source of competitive advantage for the firm (Martín de Castro *et al.*, 2007). This initial research effort has its main logic in the well-known statement from Kaplan and Norton (1992): 'what you cannot measure, you cannot manage in an efficient way'.

One of the main tasks for effective knowledge management is to analyse which intangible assets can become critical for organizational success. In general competition among firms, and particularly in knowledge-based industries, strategic management of intellectual capital has a greater relevance to competitive success than the strategic allocation of physical and financial resources (Tseng and Goo, 2005).

Nevertheless, conventional measures and instruments commonly fail when attempting to determine value creation at corporate or business levels in the knowledge-based economy. Traditional performance measures are thus insufficient when addressing strategic decision-making within the firm, and when managers face different investment alternatives. With the traditional accounting instruments, in which the value-creation potential of these investments appear on the income statement as costs (Roos, 1998), attempting to realize the true value of intellectual capital becomes critical (Tseng and Goo, 2005).

Reed *et al.* (2006) point out that the intellectual capital perspective is focused on the stocks and flows of knowledge capital embedded in an organization, and is posited to have a direct association with the firm's financial performance. Nevertheless, in this line of work, a cross-sectional or static perspective is adopted, attempting to 'take a picture' of the intangible factors of the firm, which otherwise remain 'invisible', but which are responsible to a great extent for the firm's success. Intellectual capital is therefore an attempt to carry out an inventory of intangible resources, by analysing the knowledge base that is available to the firm. This makes it necessary to choose a certain moment in time at which to perform the analysis of the firm's knowledge stocks, though comparison among different periods or among competitors can provide highly valuable information about the stock of knowledge or competitive advantages.

To summarize, this stream comes originally from the business and professional domain, and tries to identify, classify and value the different stocks of knowledge that organizations can own or control, and that usually show a highly heterogeneous nature. This represents an important advance towards knowing what occurs in the 'black box' between intellectual capital and corporate value (Tseng and Goo, 2005).

The goal of the intellectual capital-based view of competitive advantage is therefore to determine how intellectual capital assets lead to an increase in corporate value through superior competitive positions. As will be shown in the following subsections, this is not only interesting for the empirical problems that attach to the knowledge-based view of the firm, and the resource-based view of competitive advantage, but also for knowledge managers interested in how to extract maximum value from the resources available, and how they are deployed.

Characteristics and components

In later chapters of this book, intellectual capital will be analysed in detail. Nevertheless, in this subsection, a brief comment is provided about the main characteristics of this construct, and about which components or building blocks it includes according to the extant literature.

The work of Tseng and Goo (2005) highlights that intellectual capital is a broad-based term usually considered as synonymous with a firm's intangible assets. According to these authors, this construct holds particular characteristics that can be summarized as its core nature; its effect on firm performance; and its economic properties.

As we mentioned in the previous definition, intellectual capital has an intangible core nature. This means that intellectual capital assets are invisible and intangible, and thus pose an important problem for traditional accounting measures (Roos, 1998).

The effects of intellectual capital assets on a firm's performance usually show time delays (Tseng and Goo, 2005). Thus, developing these kinds of assets not only requires the allocation of resources and investments, but also takes some time to fully implement them as elements of working capital. This kind of inertia or delay can easily be connected to the path-dependence dynamics that have already been explained, when describing the main phenomena considered by the knowledge-based view of the firm.

The main economic properties of intellectual capital are related to the 'law of increasing returns'. While traditional production factors (namely land, capital and labour) all follow the law of decreasing returns, knowledge and knowledge-based assets, in contrast, enjoy increasing returns (Tseng and Goo, 2005), as the Verdoorn law states for productivity in the presence of important scale economies on the basis of economic returns with circular and cumulative causes. This makes it necessary to use a multiplication rule, rather than the addition rule usually employed in conventional financial statements devoted to physical assets.

There are several assets with very different natures that meet the above-mentioned characteristics, but which are developed and reinforced through different means. Intellectual capital is not monolithic, but a multifaceted reality including a wide range of components that need different decisions to be made. For this reason, there are several proposals dealing with identifying, measuring, and in some cases also managing, the intellectual capital of organizations.

While there is no general agreement about the main structure or elements and integrative variables that should be considered intellectual capital, a three-faceted framework is usually mentioned, considering human capital, structural (internal) capital and relationship (external) capital (Tseng and Goo, 2005), or human, organizational and social capital (Subramaniam and Youndt, 2005).

These different kinds of intellectual capital enable organizations either to reinforce or to transform their prevailing knowledge (Subramaniam and Youndt, 2005), but nevertheless requiring unique kinds of investment: human capital requiring the hiring, training and retaining of employees; organizational capital requiring the establishment of knowledge-storage devices and structured recurrent practices; and social capital requiring the development of norms that facilitate interactions, relationships and collaboration.

Providing a theory for intellectual capital

Roos (1998) noted that intellectual capital was originally conceived by John Kenneth Galbraith in the late 1960s, but this concept has become popular only recently. Nevertheless, this pioneer of the intellectual capital stream of research describes quite well the origins of this perspective with these words of an executive: 'Whereas knowledge management is a theory in search of practice, intellectual capital is a practice in search of a theory.'

The origins of the intellectual capital movement can be traced to the practitioners interested in measuring intangible assets. This measurement was focused on developing a new reporting mechanism to enable non-financial, qualitative items of intellectual capital to be measured alongside traditional quantifiable financial data (Tseng and Goo, 2005).

The language of intellectual capital helps managers to see an existing situation in a way they have not seen it before (Roos, 1998). Nevertheless, management scholars are trying to link intellectual capital with existing strategic management theory, to overcome the main empirical problems that the knowledge and resource-based theories indicate.

Providing a theoretical rationale for intellectual capital as a complement to the measurement stream aims at improving the visualization of company value creation to enable comprehensive management of intellectual capital (Tseng and Goo, 2005). Thus the intellectual capital perspective has since been adopted by academics as a useful framework for describing firm resources, and value creation, allowing the empirical testing of the knowledge-based view of the firm and the resource-based view of competitive advantage. With a common perspective and terminology emerging from case material and practitioner experience, and the connections to the strategic management literature, the practical applications and pragmatic approach of the intellectual capital-based view of competitive advantage will become a vey powerful tool for both managers and scholars.

Intellectual capital and the knowledge-based view of the firm

One of the main contributions of the intellectual capital perspective is providing the terminology and measurement tools needed in order to reduce the gap between practitioners and academics when fully developing the knowledge-based view of the firm.

According to Roos (1998), the concept of intellectual capital enables a refinement of the resource-based view of the firm, connecting it to the resource-based view of competitive advantage, and to the intangible assets measured by intellectual capital. This author suggested considering the resource-based view as some form of the knowledge-based view of the firm, and placing intellectual capital as the defined 'resource' of the resource-based view, using the basic assumptions of resource heterogeneity and immobility.

Roos (1998) also claimed, for the utility of the concept of intellectual capital, in order to express the scale and scope of firms, another critical issue addressed by the knowledge-based view of the firm. Thus, considering intellectual capital as something between the conventional distinction between product and knowledge domains, the boundaries of a certain firm could be depicted according to the potential intellectual capital it can access and put to use.

The principles of the law of increasing returns are not only a key element for managing intellectual capital within firms, but also represent one of the essential arguments for those new theories of the firm seeking to explain how the knowledge-based economy works. As Roos (1998) pointed out, it is increasingly questioned whether the application of neo-classical economics is the best way to describe what is going on in the world economy, in particular because a growing share of the world economy is better described by the law of increasing returns working in parallel with the law of decreasing returns.

Intellectual capital and the resource-based view of competitive advantage

According to the resource-based view, the value-creating capability of an organization comes not from the dynamics of the industry, but rather from idiosyncratic endowments of resources that generate firm heterogeneity, and differences in organizational returns, which provide the basics for sustainable competitive advantage.

Nevertheless, while this theoretically convincing theory of competitive advantage examines the nature and quality of resources deployed in the value creation process, it does not explain the resource deployment process and how the resulting value is created. Thus the relationships between resources (input) and corporate value (output) is assumed, but not explained (Tseng and Goo, 2005). Probably the main reason for this missing link between resources and value creation is a problem of measurement, because, most of the time, competitive advantages are based on firm-specific interaction of resources, which themselves are intangible, and therefore unobservable (Reed *et al.*, 2006).

Critics of the resource-based view (Priem and Butler, 2001; Foss and Knudsen, 2003) have raised questions about its legitimacy as a theory, precisely because it is extremely difficult to parameterize and to test empirically its main axioms without measuring intangible assets in some detail. Reed *et al.* (2006) consider that the empirical testing problems of the resource-based view are all related to its lack of specificity. However, they propose a pragmatic, though partial, resolution from a mid-range theory that they call 'an intellectual capital-based view of the firm'. Their main contribution is that, as a mid-range theory, the intellectual capital perspective enables a better hypotheses development and empirical testing for the resource-based view, adopting a partial rather than a generalized view in this research.

According to these authors, the intellectual capital-based view of competitive advantage allows academics to overcome five major concerns about the resource-based view:

(i) It is not prescriptive and does not explain to the executives what specific resources should be developed in order to obtain a competitive advantage. As a mid-range theory, the intellectual capital perspective deals with different kinds of assets, classified as people, social relations, and information technology systems and processes. These specific aspects of the more general approach of the resource-based view can be connected more effectively to competitive advantage and value creation.

(ii) It lacks a clear definition of competitive advantage, but the intellectual capital perspective can depict the resource characteristics that allow a firm to outperform rivals in the same industry, providing, therefore, practical and specific definitions of competitive advantage.

(iii) It is tautological, because resources are defined in terms of the performance outcome associated with them. While this problem has provided an interesting debate and subsequent advance for the resource-based view (Barney, 2001; Priem and Butler, 2001; Peteraf and Barney, 2003), the intellectual capital perspective can overtake it by defining knowledge resources by their theoretical associations with competitive advantage and not by their empirical financial association (Reed *et al.*, 2006).

(iv) Its relevant domain is ambiguous. Nevertheless, the intellectual-capital-based view of competitive advantage, as a mid-range theory, examines intra-industry intangible resources association with financial performance. Holding constant exogenous influences to be industry or geographical effects, this approach can focus effectively on competitive advantages related to the resource-based level.

(v) It is too general, so there can be several advantageous resource configurations, thus suggesting equifinality. Addressing the empirical test at the resource-based level, as the intellectual capital perspective can do, and holding constant exogenous and other endogenous effects, can also solve this problem.

As has been explained, the intellectual capital-based view of competitive advantage has experienced different stages in its development, and all along has been more integrated into strategic management theory, as was outlined in the previous subsection. Nevertheless, at present there are not only important solutions for other wider-ranging theories, but also important challenges that this perspective has to deal with.

The 'first stage' of this research stream was based strictly on managerial practice, and its focus was on everything done to identify, describe and measure intellectual capital. Roos (1998) included in this stage the pioneering ideas of the different intellectual capital models presented in the 1990s and the beginnings of the twenty-first century (see Chapter 2 for further information about these models).

Once intellectual capital and its main components have been defined, and the relevance of this construct for sustainable competitive advantage has been recognized by both practitioners and academics (Petty and Guthrie, 2000), the 'second stage' of development in intellectual capital research is needed, whereby empirical tests legitimize the study of this construct and provide more robust evidence on which to build (Reed *et al.*, 2006). This is where the connections with the resource-based view of competitive advantage become critical, so the intellectual capital-based view of competitive advantage can act as a mid-range theory for hypothesizing and testing empirically the influence of the different elements of intellectual capital on competitive advantage.

To move forward, the intellectual capital-based view of competitive advantage needs to address two important challenges: consolidate measures for a better time or inter-firm comparison; and test the arguments of the resource-based view partially in different contexts. On the one hand, as Roos (1998) pointed out, an additional step is to consolidate measures and focus on changes in intellectual capital, moving from an intellectual capital or knowledge stock approach towards one employing organizational learning or knowledge flows. For dynamic or firm comparisons, improving the intellectual capital indexes can provide a wealth of insights that complement traditional efficiency measures. On the other hand, Reed *et al.* (2006) called for a 'third stage' of development in intellectual capital research, whereby empirical tests can further explore intellectual capital in different competitive contexts, with the objective of expanding the scope of the mid-range contingency expectations presented within.

Intellectual capital and firm competition

The results obtained by recent contributions of the intellectual capital approach provide two important recommendations for those executives who are in charge of managing intellectual capital in their organizations: the different elements of intellectual capital are usually intertwined, so it is necessary to adopt a holistic perspective when managing these kinds of assets; and innovation-based competitive advantages are closely tied to the successful deployment of a firm's intellectual capital.

According to the resource-based view, complementary resources are more likely to contribute to firms attaining and sustaining superior performance precisely because of their combination and integration. The importance of resource integration for obtaining higher rents is based on the rationale that the combined set is indivisible, and therefore distinctive (Reed *et al.*, 2006). In fact, and in the same vein, O'Donnell and O'Regan (2000), Ulrich (1998) and Tseng and Goo (2005) provide empirical support for the positive relationship and intertwined action of the different components of intellectual capital, which complement one another to increase corporate value, rather than acting separately and independently.

The results that Reed *et al.* (2006) suggested indicate that intellectual capital interactions are best understood within the very specific industry conditions in which they are developed. They found evidence of a reinforcing or enhancing effect of some components over others, but also that intellectual capital interactions in some markets may experience diminished returns (that is, too much of a good thing is not always good). This makes it necessary to carry out specific empirical tests, because the context matters, and, as noted already, knowledge resources are best understood within the specific context in which they are developed.

Subramaniam and Youndt (2005) also proposed that inherent differences in the key attributes of intellectual capital components cause each of them to have a particular reinforcing or transforming influence on knowledge, showing intertwined action in organizations. Nevertheless, their work was focused on the effects of the different components of intellectual capital, either independently or through their inter-relationships, reinforcing or transforming organizational knowledge to develop incremental and radical innovative capabilities.

While the relationship between intellectual capital and innovation is the main purpose of the empirical research to be presented in this book, it is important to remark here that there is a growing tendency among studies examining innovation to use knowledge or intellectual capital as antecedents, and studies investigating knowledge and intellectual capital frequently use innovation as an outcome (Subramaniam and Youndt, 2005).

The interest of the intellectual capital-based view of competitive advantage in connecting the intangible inputs and visible outputs of the resource-based view can discover a very fruitful field in innovation-based competitive advantage. Thus managers must pay attention not only to the final markets but also to the factor markets, trying to keep the balance between knowledge exploration and exploitation in order to reach ongoing innovations for sustaining competitive advantage. The following subsections will comment on these two ideas.

Goods and services market, and intellectual capital markets

We have seen that firms compete against each other with proposals of value creation for their customer according to the industry in which they are active. Then, according to industrial organization economics, we could think that the only decision a company has to make is about which industry shows the best commercial and profitability margins, and then, once located in it, harvests the best results. This logic would fit perfectly with the focus of competitive forces (Porter, 1980), according to which, firm decisions should be taken as indicated in the following outline: (i) activity field is chosen according to the structural attractiveness of each industry;(ii) entrance strategies are formulated according to competitors' rational decisions; and(iii) if the firm lacks any productive factors that might be needed, these are acquired or developed. However, this intuitive argumental sequence becomes complicated at the third stage.

What happens when, in order to complete successfully within a certain knowledge-intensive sector, with an outstanding potential attractiveness, our firm needs essential competitive factors such as a good reputation, organizational culture and highly qualified employees? If we analyse this simple example in some detail we realize that this company, to enter the desired sector, needs relational capital to maintain an appropriate relationship with its customers (the reputation, organizational image or trademark); structural capital (or what we have previously called collective or social knowledge) in order to foster communication and co-ordination through a shared social identity; and human capital, or individual specialized knowledge. The firm in this example needs knowledge of different kinds as an essential production factor for the industry. But how can it get this knowledge? Specialized individual knowledge can be acquired by hiring expert workers through work markets, though often this is neither easy nor cheap. But does a market for organizational culture exist? Does a market for corporate reputation exist? Can we buy these kinds of knowledge-based assets?

As we can see, the problem of acquiring the necessary resources to be competitive seems even more complicated than that of choosing a market in which to operate, facing rival firms in the sale of the finished product. Thus firms will also have to compete against each other to obtain the best production factors that could distinguish them from the rest, allowing them to design a proposal with more value added for theirs customers. Now it is necessary to change the point of view used to pose an appropriate sequence for taking business decisions. We shall direct our attention away from the external, market or demand perspective and towards an internal and production-factors-orientated one.

Thus, the resource-based view becomes considerably relevant (Wernerfelt, 1984; Barney, 1991; Grant, 1991; Peteraf, 1993). From it, and also in a parallel way, the knowledge-based theory of the firm has been developed, as well as the approaches of knowledge creation and transfer. Although there has been some controversy about whether the resource-based view can be considered a full theory (Barney, 2001; Priem and Butler, 2001), it is undeniable that this theoretical approach has led to an important change in the main perspective adopted for researching strategic analysis during recent years.

According to Teece *et al.* (1997), from the resource-based point of view, decision-making in the firm follows this path: (i) the organization identifies the resources that it considers to be strategic keys; (ii) it analyses the markets in which those resources will generate more rents; and (iii) it proceeds to the most effective use of the rents generated by those assets by means of their integration into related markets, selling the intermediate product to related companies, or selling those assets directly to other firms. Therefore the resource-based view is focused on the rents that the owners receive from scarce and firm-specific resources, instead of analysing the economic benefits (profits) that come from the competitive position of the firm's products in the market.

The change in the logic for analysis is evident, moving from a completely exogenous point of view towards a more endogenous focus. In the words of Fernández and Suárez (1996: 76), the resource-based view moves the focus to firm production factors because it recognizes that building sustainable competitive advantages that are able to provide long-term rents depend mainly on those internal factors.

The dynamic capabilities approach analyses why some companies are able to build a competitive advantage in environments with rapid changes, as in the case of the present knowledge-based economy, as we have already explained. Teece *et al.* (1997) affirmed that this approach is very relevant in a Schumpeterian world, where competition is based mainly on innovation, price and outcomes rivalry, growing returns and the 'creative destruction' that generate the organizational capabilities available to the firm.

If controlling scarce resources is a source of economic profitability, then through the combination of those resources and shaping capabilities, some issues, such as expertise acquisition, knowledge management, know-how and learning become topics of outstanding strategic relevance (Teece *et al.* 1997).

The competitive advantage of the firm lies in the organizational and managerial processes that come from its specific resources and the different alternatives they pose. Organizational and managerial processes are the way in which things are done in the firm, or its organizational routines, or their current practical patterns and learning that are applied for developing organizational capabilities. The current resource position is the specific endowment of technology, intellectual property, asset complementary, customer base, and the company's relationships with suppliers and other external agents.

In summary, resource- and capability-based approaches suggest that firms must take strategic positions according to those resources and organizational capabilities that demonstrate being unique, valuable and inimitable, rather than starting their strategy from the delivered products and services that arise from *applying* those capabilities, because a resource- or capability-based competitive advantage offers higher possibilities for its maintenance than one that is only based on product and market positioning (Zack, 1999).

Barney (1991) settled on basic axioms of the resource-based view as: (i) the firms of a certain industry (or strategic group) can be heterogeneous in relation to the strategic resources they control; and (ii) these resources cannot be perfectly mobile across companies, and therefore heterogeneity among companies can be sustainable and durable.

As well as these axioms, it is necessary to add the important role that historical dependence plays in the process of accumulating resources and organizational capabilities. According to Teece *et al.* (1997), the current position of a company, in relation to its endowment of resources and capabilities, is determined by a path followed from the past. This notion of path dependence recognizes the importance of company history. Thus the previous investments of the firm, as well as its range of routines (its 'history') determine its future behaviour. The importance of path dependence is enhanced when we are under conditions of growing returns. Growing returns have multiple sources, such as the presence of complementary assets and support infrastructures, learning by doing, and scale economies in production and distribution activities.

When considering resources and capabilities as the main determinants of a firm's economic results, it becomes indispensable to analyse these internal elements in order to carry out the strategic management process of the firm. As Navas and Guerras (2002) said, the purpose of resource and capabilities analysis is to identify the potential that the firm has to establish competitive advantages based on the already available set of resources and abilities, or even from those it will be able to access in the future.

Exploration and exploitation

From all that has been said up to this point, we can conclude that a successful firm is one that is able to manage its intellectual capital efficiently. Thus managers in charge of intangible assets must assess the alternatives for the acquisition and utilization of existing knowledge and/or the creation of new knowledge, with the intention of improving economic outcomes (Boerner *et al.*, 2001).

Efficient knowledge management about intellectual capital assets that involves organizational learning is also an organizational meta-capability to foster knowledge creation and acquisition, and to disseminate it throughout the whole organization, incorporating this knowledge into the firm's products, services and systems (Nonaka and Takeuchi, 1995). Crossan *et al.* (1999) considered that organizational learning is one of the main means of encouraging a strategic renewal in the company. This renewal harmonizes continuity and change at an organizational level, and allows a firm to adapt to the hyper-speed rhythms of change present in the knowledge economy environment. Strategic renewal through organizational learning demands that the firm explores and learns new paths, while exploiting what it has already learnt. Thus firms must perform different activities.

Bearing in mind the previous ideas discussed, two different groups of activities that appear in the process of managing intellectual capital can be considered. These two kinds of actions can be employed by the firm in order to obtain and preserve competitive advantages over their rivals based directly on knowledge. The first of these groups of activities is linked to the creation and acquisition of new knowledge, while the second is closely related to the use, incorporation and dissemination of already available knowledge. As it can easily be connected to the resource-based view, the former set of activities is related to obtaining competitive advantages, while the latter pursues the reinforcing and sustaining of them.

When a firm adopts a strategic focus for knowledge management in order to achieve competitive advantages based on organizational learning, it must recognize and manage the tension between exploration and exploitation, which represents the key challenge for strategic renewal. It is one of the central ideas of the work of Crossan *et al.* (1999) that organizational learning involves an important tension between the assimilation of new learning (exploration) and the use of what has already been learnt (exploitation). On the one hand, exploration will seek innovation and novelty mainly through the generation of new knowledge, while on the other, exploitation will be focused on applying already-generated knowledge, relying to a wider extent on knowledge transfer to incorporate this knowledge into different areas (products, customers or geographical areas).

Bierly and Chakrabarti (1996) extended this typology of generic strategies for knowledge management even further. According to them, there are several strategic decisions that managers must make in order to configure and manage the intellectual capital and organizational learning in their firms. Key decisions must be stated in an explicit way by the top managers, or they can be expressed implicitly through behaviour regarding resource allocation, and posing strategic goals and incentives to align employees with those objectives. From the four fundamental areas that Bierly and Chakrabarti (1996) proposed for these decisions, two can be specially highlighted for our purpose: sources of learning or the focus of intellectual capital; and the trade-off decision about deep/wide learning or intellectual capital specialization/variety.

The decisions about sources of learning or intellectual capital focus are related to the relevance that the firm confers on internal and external knowledge. On the one hand, an internal focus appears when the members of the organization generate, share and distribute knowledge within the company, while, on the other hand, organizations with an external focus seek knowledge from outside the firm in order to incorporate it through acquisition or imitation, though subsequently this captured knowledge is transferred throughout the whole organization.

Broadly speaking, an external focus is usually tied to exploration activities, while the internal focus can be related to exploitation activities. We must also take into account that acquiring tacit knowledge from another firm can be difficult, or even impossible, and this may force a firm to choose internal learning as the only means of intellectual capital accumulation.

The decisions about deep/wide learning or intellectual capital specialization/variety are those in which managers determine how wide the knowledge base must be that the firm wants to exploit to implement its strategy, and how specialized each element of intellectual capital must be. In this sense, the more limited the available resources, the more appropriate it is to be focused on exploiting a limited domain of knowledge (or core competences, using the terms of Hamel and Prahalad, 1994). However, when the organization has a wide knowledge base and different intellectual capital assets, this allows for several ways of achieving a better competitive position, in particular when we take into account that sustaining competitive advantage can depend strongly on how complex is the combination of the specialized knowledge pieces that appears in the organizational knowledge base (Reed and DeFillipi, 1990).

This is the main decision when determining if an organization will focus on the exploration of new knowledge territories (the radical learning focus) or mainly be involved in the exploitation of already available knowledge (incremental learning). Incremental learning can be more effective for the firm in the short term, but a certain level of radical learning is always necessary to achieve long-term survival (March, 1991), and demands highly specialized assets. Nevertheless, the costs of knowledge exploration, with a wider knowledge base, are usually very high and to capitalize conveniently on them it is necessary to move into knowledge exploitation activities. It is obvious to consider that an optimum position must be found in striking a balance between knowledge exploration and exploitation. However, it is difficult for a firm to be equally skilled in both types of activities, because it faces restrictions about available resources, and exploration and exploitation require different organizational cultures and structures (Hedlund, 1994).

2
Firms' Intellectual Capital

Firms' intellectual capital: origins and concept

The rise of the knowledge-based economy and society has been attributed to the predominance of intellectual capital (IC) as a key resource for obtaining sustained competitive advantages among firms (Bontis, 2001; Dean and Kretschmer, 2007).

While it has been recognized that economic wealth comes from knowledge assets – or intellectual capital – and its useful application (Teece, 1998), the emphasis on it is relatively new, and the management of firms' intellectual capital has become one of the key tasks on the executive agenda. Nevertheless, this work is particularly difficult because of the problems involved in its identification, measurement and strategic assessment. In this situation, models of intellectual capital become highly relevant, because they not only allow us to understand the nature of these assets, but also to carry out their measurement.

However, if the nature of intellectual capital assets is both non-financial and intangible, how can it be identified, measured and managed? Financial statements are inadequate and insufficient when a firm's value lies primarily in its intangible assets (Moon and Kym, 2006).

The term 'intellectual capital' is used as a synonym for intangible or knowledge assets (Stewart, 1991). Calling it 'capital' makes reference to its economic roots, because it was described in 1969 by the economist J.K. Galbraith as both a process of value creation and a bundle of assets. Reed *et al.* (2006) provide several arguments to consider the intellectual capital-based view of the firm as a pragmatic and focused theory that helps in the empirical testing of the resource-based view (RBV). In this sense, its definition as 'basic competences of intangible character that allow the creation and maintenance of competitive advantage', argues how we can tie intellectual capital to the RBV.

Teece's 'intellectual capital' represents one of the earliest and most cited developments of the term. However, Teece's concept was never defined explicitly, and his use oscillates between a narrow financial use and a general equation of intellectual capital with intangible resources (Dean and Kretschmer, 2007).

Some of the definitions converge at the same point, which can be known as 'the combination of intangible assets that allows the company to operate' (Brooking, 1997: 25), the difference between the market value of the company and the replacement cost of its assets (Bontis, 1996), or as the sum of all knowledge possessed by the employees of an organization that confer on it a competitive advantage. In short, intellectual material – 'knowledge, information, intellectual property, expertise ... which can be used to create wealth' (Stewart, 1998: 9–10).

As can be seen in some of these definitions, intellectual capital is the stock or fund of knowledge, intangible assets, and ultimately intangible resources and capabilities, that allows the development of basic business processes of organizations, and enables the achievement of competitive advantage. This potentially strategic nature of intellectual capital is embodied in the following definition translated from Bueno (1998: 221): 'a set of competences, invisible or intangible, allowing the firm to operate, creating value for it'.

It is difficult to provide a unified definition of intellectual capital, and even more difficult to propose a commonly accepted typology for it, because this phenomenon is still at an emerging stage of development. Table 2.1 summarizes some of the definitions of intellectual capital. As it can be appreciated from it, the term has not yet solidified, and its identification with the concept of 'capital' –in accountability terms – presents controversy (Dean and Kretschmer, 2007).

Nevertheless, intellectual capital demands a more rigorous theoretical framework (Cabrita and Bontis, 2008). In this regard, since the beginning of the 1990s, many proposals have been raised around its theoretical identification and classification.

From a strategic point of view, it is linked to the ability to create and apply the potential of an organization's knowledge base. In essence, three key characteristics of this construct can be described as follows:

(i) Its intangibility
(ii) Its potential to create value
(iii) The growth effect of collective practice and synergies

Table 2.1 Intellectual capital concepts

Concept	Authors
The difference between an organization's market value and book value	Galbraith (1969)
The difference between the market value of the company and the replacement cost of assets	Bontis (1996)
The combination of market assets, human-centred assets, intellectual property assets and infrastructure assets	Brooking (1996)
The gap between market and book value of the firm	Sveiby (1997)
The gap between a firm's market value and its financial capital (book value of a firm's equity)	Edvinsson and Malone (1997)
Knowledge, information, intellectual property, expertise that can be used to create wealth	Stewart (1997)
Set of intangibles, invisible or intangible, off-balance, allowing the company to operate, creating value for it	Bueno (1998)
Knowledge and knowing capability of a social collectivity	Nahapiet and Ghoshal (1998)
Essentially comprises all immaterial resources that could be considered as assets, being possible to acquire, combine, transform and exploit, and to which it is possible to assign, in principle, a capitalized value	Granstrand (1999)
Intellectual assets, knowledge assets, total stock of knowledge-based equity possessed by a firm	Dzinkowski (2000)
Includes knowledge, competence and intellectual property. Also includes other intangibles such as brands, reputation and customer relationships	Teece (2000)
Represents the stock of knowledge existing in an organization at a particular point in time	Bontis *et al.* (2002)
Set of intangibles, invisible or intangible, off-balance, allowing the company to operate, creating value for it. Includes human, technological, organizational, relational and social capitals	CIC (2003)
The sum of all knowledge that firms utilize for competitive advantage	Subramanian and Youndt (2005)
Includes the intangible assets of an organization that are not recorded in financial statements but which may constitute 80 per cent of the market value of the organization	Martínez-Torres (2006)

(Continued)

Table 2.1 Continued

Concept	Authors
Basic competences of intangible character that allow the creation and maintenance of competitive advantage	Reed *et al.* (2006)
Set of intangible resources and capabilities possessed or controlled by a firm	Alama Salazar (2008)
The knowledge assets that can be converted into value. A matter of creating and supporting connectivity between sets of expertise, experience and competences, both inside and outside the organization	Cabrita and Bontis (2008)
Represents knowledge-related intangible assets embedded in an organization	Chang *et al.* (2008)
The total capabilities, knowledge, culture, strategy, process, intellectual property and relational networks of a company that create value or competitive advantage and help a company to achieve its goals	Hsu and Fang (2009)

In this sense, and based on a wide literature review, Dean and Kretschmer (2007) identify other characteristics of intellectual capital:

- Weightless
- Tradable
- Cheap to reproduce
- Appreciates rather than depreciates with use
- Multiple, simultaneous applications
- Effective interface among information technology, business development and human resources
- Inexhaustible: ability to leverage knowledge capital is unlimited
- Socially and contextually embedded knowledge
- Closely related to social capital
- Dominating as a means of production
- Ownership is central
- Fixed or flexible, both the input and output of the value creation process
- Transfer cost hard to calibrate
- Property rights are limited in many of these assets

Both Brennan and Connell (2000) and Guthrie *et al.* (2003) provided an understanding of the evolution of the intellectual capital field from its roots.

From the practitioner point of view, one of the first and best-known efforts in developing and implementing a measurement model of intellectual capital was made in the 1980s at Skandia, a Swedish financial services company. Edvinsson and Malone (1997: 11) defined intellectual capital as 'those dimensions beyond the human capital that were left behind when the staff went home'.

Nevertheless, the need to adapt theoretical and empirical models to different industry contexts as well as to new social and economic trends justifies an effort in improving previous proposals. In this setting, empirical evidence for the classification and measurement of intellectual capital is still necessary and empirically supported models are scarce (Lim and Dallimore, 2004; Martínez-Torres, 2006).

The different types of intellectual capital represent different kinds of intangible resources and capabilities. Nevertheless, in spite of their strategic nature, these assets would not all have the same value for a firm (Itami and Roehl, 1987; Aaker, 1989; Prahalad and Hamel, 1990; Hall, 1992, 1993) emphasizing the importance of certain intangibles. Thus these kinds of differences can be considered useful steps for strategic management. They can help in making decisions about the actions the firm must perform, and about the implementation of programmes that allow it to protect, maintain or develop its most valuable intangible assets. Thus, before exploring the relationship between any specific kind of intellectual asset and competitive advantage, a clear identification of the main components of intellectual capital is required.

As we have noted previously, in an effort to understand this new competitive dynamic developing since the mid-1990s, numerous theoretical proposals about the concept and measurement of firm's intellectual capital have emerged (Kaplan and Norton, 1992; Bontis, 1996; Brooking, 1996; Edvinsson and Malone, 1997; Sveiby, 1997; CIC, 2003, among others).

But, what are the different components of intellectual capital? Edvinsson and Malone (1997) proposed that it is a two-level construct: human capital (the knowledge created by, and stored in a firm's employees); and structural capital (the embodiment, empowerment and supportive infrastructure of human capital). They then divide structural capital into organizational capital (knowledge created by and stored in a firm's information technology systems and processes, that speeds the knowledge flow through the organization) and customer capital (the relationships a firm has with its customers).

Based on a literature review, at international level there is a general agreement about the three main components of intellectual capital (Bontis *et al.*, 2000; Martínez-Torres, 2006). In a wide sense, these represent different expressions of firms' knowledge stocks. This triple nature of intellectual assets is being revisited by different lines of research trying to reconcile the concept of intellectual capital (Dean and Kretschmer, 2007).

In this way, several contributions have provided different frameworks for classifying the components of intellectual capital, as well as for establishing a series of indicators for its measurement. Thus, according to the literature review, in a first step, three main components can be found: (i) human capital; (ii) structural capital; and (iii) customer or relational capital (Subramanian and Youndt, 2005; Martínez-Torres, 2006: Hsu and Fang, 2009). See Figure 2.1.

- *Human capital* refers to the tacit or explicit knowledge employees possess, as well as to their ability to generate it, which is useful for the firm, and includes values and attitudes, aptitudes and know-how.
- *Structural capital* includes technological and organizational capital. The first refers to the combination of knowledge directly linked to the development of the activities and functions of the technical system of the organization, responsible for obtaining products and services; while the second can be seen as the combination of explicit and implicit, formal and informal knowledge which, in an effective and efficient way, structures and develops the organizational activity of the firm. This includes culture – implicit and informal knowledge, structure – explicit and formal knowledge, and organizational learning – implicit and explicit, formal and informal knowledge renewal processes).
- *Relational capital* refers to the value to the organization of the relationships it maintains with the main agents connected with its basic business processes – customers, suppliers, allies and so on, as

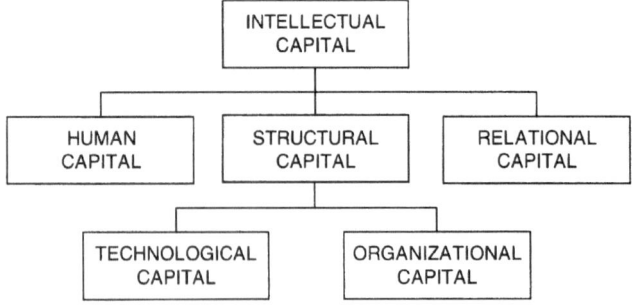

Figure 2.1   Components of intellectual capital

well as the value to the organization of the relationships it maintains with other social agents and its surroundings.

Nevertheless, empirical evidence is needed to determine the level of aggregation that intellectual capital components must adopt in practice. This is one of the purposes of this work: to find out the main components or building blocks of an intellectual capital balance sheet in the case of manufacturing firms.

Brooking (1996) highlighted the differences between intellectual property assets – focused on technological knowledge – and infrastructure assets – focused on organizational knowledge, and gave a broader concept for market assets,including customer assets.

As can be seen from the previous definition of structural capital, because of its heterogeneous nature, it can be divided into two types of capital: technological and organizational. In the same way, relational capital can be divided into business and social capital. This more detailed classification may allow a better understanding of these types of organizational factor. In this sense, the Intellectus Model (CIC, 2003) is a good example of the theoretical proposals regarding intellectual capital that are becoming more complex and detailed every day. This encourages analytical reflection among managers and senior information officers, but it can also be seen as an extensive proliferation of criteria and categories of intangible assets.

Bearing this aim in mind, in this research we take the three most common components of intellectual capital (namely human capital, structural capital and relational capital) and test empirically to discover whether this grouping of intangible assets is supported by the evidence obtained from a sample of knowledge-intensive firms.

As we can see from Table 2.2, there have been many contributions on intellectual capital typologies, using different terminologies, and in the case of relational assets, focusing more-or-less on the customer relationships. Nevertheless, as we highlighted previously, certain agreement can be found among researchers about the three main blocks of intellectual capital.

The following subsections will present a deeper analysis of the human, structural (including technological and organizational) and relational assets, offering definitions as well as dimensions and variables.

Human capital

From a macroeconomic point of view, the OECD (1992) recognized human capital as a primary driver of competitiveness, prosperity

Table 2.2 Intellectual capital typology

Authors	Individual perspective	Intellectual capital blocks	
		Collective or organizational perspective	
		Internal organizational perspective	External organizational perspective
Kaplan and Norton (1992)	Learning and growth	Internal processes	Customers
Saint Onge (1996)	Human capital	Structural capital	Customer capital
Brooking (1996)	Assets based on the individuals	Assets based on intellectual property Assets based on infrastructure	Market assets
Sveiby (1997)	Competences	Internal structure	External structure
Edvinsson (1997)	Human framework	Processes framework	Customer framework
Edvinsson and Malone (1997)	Human capital	Structural capital	Customer capital
Bontis (1998)	Human capital	Structural capital	Customer capital
Euroforum (1998)	Human capital	Structural capital	Relational capital
Nahapiet and Ghoshal (1998)	Individual explicit knowledge Individual tacit knowledge	Social explicit knowledge Social implicit knowledge	
Dzinkowski (2000)	Human capital	Organizational capital	Customer capital
McElroy (2002)	Human capital	Structural capital Innovation capital Process capital	Social capital Intra-social capital Inter-social capital
CIC (2003)	Human capital	Technological capital Organizational capital	Business capital Social capital

Guthrie et al. (2004)	Human capital	Internal capital	External capital
Bueno et al. (2004)	Human capital	Organizational capital Technological capital	Business relational capital Social capital
Chen et al. (2004)	Human capital	Innovation capital Structural capital	Customer capital
Joia (2004)	Human capital	Internal capital Structural capital Innovation capital	External capital
Ordoñez de Pablos (2004)	Human capital	Structural capital Technological capital Organizational capital	Relational capital
Subramanian and Youndt (2005)	Human capital Internal social capital	Organizational capital	External social capital
Reed et al. (2006)	Human capital	Structural capital	Relational capital
Martínez-Torres (2006)	Human capital	Structural capital	Relational capital
Alama Salazar (2008)	Human capital	Structural capital	Relational capital
Cabrita and Bontis (2008)	Human capital	Structural capital	Relational capital
Chang et al. (2008)	Human capital	Structural capital	Relational capital
Hsu and Fang (2009)	Human capital	Structural capital	Relational capital

and economic wealth. Thus it can be considered the key element of intellectual capital and one of the most important sources of a firm's sustainable competitive advantage (Nonaka and Takeuchi, 1995; Edvinsson and Malone, 1997; Cabrita and Bontis, 2008).

Human capital refers to the knowledge – both explicit and tacit – that people possess, as well as their ability to generate it, which is useful for the operation of the organization (CIC, 2003). Table 2.3 shows a literature review of human capital definitions made by academics and practitioners.

Edvinsson and Malone (1997) remarked that human capital includes knowledge, skills, innovativeness and the ability to meet the task at hand, and noted a key characteristic: human capital cannot be owned by the company. Its strategic and managerial implications for managers are remarkable.

To advance in the understanding of the nature of human capital, it is necessary to analyse its internal structure, providing coherence for it. In this way, and based on the literature review, we could consider three main dimensions:

- *Knowledge*: refers to the knowledge employees have to carry out their tasks successfully. Includes the following variables: (i) formal education; (ii) specific training; (iii) experience: and (iv) personal development.
- *Abilities*: refers to the type of knowledge related to 'the way of doing things' (know-how). Specifically, it gathers all the utilities, dexterity and talent a person develops as a result of his/her experience and practice. Includes the following variables: (i) individual learning: (ii) collaboration/team work; (iii) communication (exchange of individual knowledge and know-how); and (iv) leadership.
- *Behaviours*: These represent knowledge about the incipient sources that lead individuals to complete their tasks. They include mental models, paradigms, beliefs and so on, and refer to: (i) feeling of belonging and commitment; (ii) self-motivation; (iii) job satisfaction; (iv) friendship: (v) flexibility; and (vi) creativity.

In her doctoral dissertation, Alama Salazar (2008) found three main components of human capital in the case of professional service firms: (i) experience and abilities, characterized by the personal experience required for successful performance, as well as by the personal abilities within the firm; (ii) professional development, which includes 'access to internal promotion' and incentive plans, but also includes other aspects such as job satisfaction, educational degrees or their university's

Table 2.3 Definitions of human capital

Concept	Authors
Human-centered assets are collective expertise, creative and problem-solving capability, leadership, entrepreneurial and managerial skills embodied by employees of the organization	Brooking (1996)
Defined as the combined knowledge, skill, innovativeness and ability of company's individual employees to meet the task at hand	Edvinsson and Malone (1997)
The capacity to act in a wide variety of situations to create both tangible and intangible assets	Sveiby (1997)
Skills and knowledge of our people	Stewart (1997)
Represents the value of knowledge and talent embedded in the employees, and includes values and attitudes, personal knowledge and skills, and behaviours	CIC (2003)
The knowledge, skills, and abilities residing with and utilized by individuals	Subramanian and Youndt (2005)
The knowledge, skills and so on of individuals	Martínez-Torres (2006)
Comprises the individual's education, skills, values and experiences	Cabrita and Bontis (2008)
Embraces all the skills and capabilities of the people working in an organization	Wu et al. (2008)
Denotes the tacit knowledge embedded in the minds of the employees. Employees generate intellectual capital through their competence, attitude, motivation, and intellectual agility	Chang et al. (2008)
The intangible elements that human capital includes refer, basically, to the knowledge acquired by a person, as well as other individual qualities such as loyalty, versatility or flexibility, which determine the productivity and value of the contribution of the individual to the company	Alama Salazar (2008)
Comprises all business capital embedded in employees and not owned by the organization. This capital may be taken away by employees, and includes employees' and managers' competence, experience, knowledge, skills, attitude, commitment and wisdom	Hsu and Fang (2009)

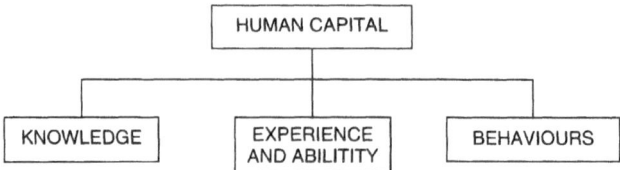

Figure 2.2   Components of human capital

reputation; and (iii) worker permanence. Permanence could be seen as the necessity to retain the outstanding experience and abilities of the most talented allies, and as a way of nurturing employees with professional development, to reinforce their skills and experience within the firm.

At this time, and based on the literature review, we can note three basic components of human capital (see Figure 2.2): (i) knowledge, embedded in the organization's employees, which might include education and training; (ii) experience and ability –the employees' know-how; and (iii) personal behaviours, willingness or attitudes, towards its task, jobs and organizations.

Structural capital

In a pragmatic and basic way, Edvinsson and Malone (1997) defined structural capital as everything that gets left behind at the office when employees go home. Nevertheless, the above conceptualization has important differences and strategic implications for this collective or organizational knowledge. Whereas human capital is possessed by the employees, making its management very difficult, structural capital is possessed, controlled and managed by the firm.

In this sense, structural capital can be seen as the skeleton and the glue of an organization, because it provides the tools and architecture for retaining, packaging, reinforcing and transferring knowledge during the business activities (Cabrita and Bontis, 2008). Table 2.4 shows the main definitions founded in the literature for the concept of structural capital.

In a similar way to human capital, and in order to advance the understanding of the nature of structural capital, we need to analyse its internal structure in a search for coherence. Nevertheless, at this point, we must note that its heterogeneous nature allows us to split this construct into two main 'capitals' or building blocks: (i) technological capital; and (ii) organizational capital.

Table 2.4 Definitions of structural capital

Concept	Authors
Includes intellectual property assets that contain the legal mechanism for protecting many corporate assets, infrastructure assets, including know-how, trade secrets, copyrights, patents, design rights, trade and service marks, as well as infrastructure assets, including those technologies, methodologies and processes that enable the organization to function, including corporate culture, databases and information and so on	Booking (1996)
The hardware, software, databases, organizational structure, patents, trademarks and everything else of organizational capability that supports employees' productivity	Edvinsson and Malone (1997)
Patents, concepts, models and computer and administrative systems	Sveiby (1997)
Stored knowledge in patents, processes, databases, networks and so on	Stewart (1997)
Technological capital. Includes know-how from research and development (R&D) activities, technological knowledge, trade secrets, intellectual property and patents	Hsieh and Tsai (2007)
Organizational capital: the institutionalized knowledge and codified experience residing within and utilized through databases, patents, manuals, systems and processes	Subramanian and Youndt (2005)
The property of the organization, such as processes, information in a database and so on	Martínez-Torres (2006)
Includes the intangible assets that form a part of the structural design of the company, which facilitate the flow of knowledge and improve the efficiency of the firm, since they make possible the integration of different functions.	Alama Salazar (2008)
Comprised of non-human assets, such as information systems, routines, procedures and databases	Cabrita and Bontis (2008)
Includes all the non-human storehouse of knowledge in organizations, including databases, organizational charts, process manuals, strategies, routines and anything whose value to the firm is higher than its material value	Wu *et al.* (2008)

(Continued)

Table 2.4 Continued

Concept	Authors
Refers to the non-human storehouses of knowledge in a firm that involve organizational structures, such as the organizational routines, the structure of the business, and various types of intellectual property	Chang *et al.* (2008)
Includes process capital and innovation capital. Process capital is defined as workflow, operation processes, specific methods, business development plans, information technology systems, co-operative culture and so on. Innovation capital is defined as intellectual property within an organization, including patents, copyrights, trademarks, know-how and so on	Hsu and Fang (2009)

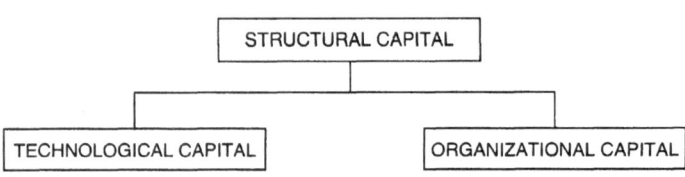

Figure 2.3 Components of structural capital

Since the start of intellectual capital studies, authors such as Brooking (1996) raised the need to differentiate those intellectual assets linked to innovation, intellectual property or technology, such as know-how, trade secrets, copyrights, patents, design rights and so on, from those related to infrastructure assets, such as corporate culture, information and telecommunication technologies, databases and so on.

In this sense, several authors, such as CIC (2003), Chen *et al.*, (2004), Joia (2004), Alama Salazar (2008) or/and Hsu and Fang (2009) have made this distinction. Figure 2.3 shows the components of structural capital.

Technological capital

As CIC (2003) pointed out, technological or innovation capital refers to the combination of organizational knowledge linked directly to the development of the activities and functions of the operation's technical systems, responsibility for obtaining new products and services, the development of efficient production processes, and the advancement of the organizational knowledge base necessary to develop future technological innovations.

Technological capital includes the following elements:

- Efforts in research and development. Typically includes research and development (R&D) expenditure; personnel linked to these efforts; and the number and relative importance of R&D projects.
- Technological infrastructure, including the purchase of technology as well as the information and telecommunications infrastructure necessary to develop technological innovations.
- Intellectual and industrial property, as the volume of legally protected and unprotected technical and scientific knowledge of the firm. Includes patents, prototypes, trade secrets, design rights, registered trademarks, licenses and so on.

Organizational capital

The second group of structural assets is linked to the organizational infrastructure (Brooking, 1996). Organizational capital results from the combination of intangible assets that are by their nature both explicit and implicit, formal as well as informal, and which, in an effective and efficient way, give structure and organizational cohesion to the different activities and business processes developed within the firm (CIC, 2003; Alama Salazar, 2008).

Organizational capital includes the following main elements:

- Organizational culture, values and attitudes. Includes the level of cultural homogeneity, or level of coherence, acceptance and general commitment to cultural values; business philosophy and ethics; social climate; or managerial commitment towards some concrete cultural values and attitudes.
- Information and telecommunications capability. Refers to the firm's ability, commitment and effective use of information and telecommunications technologies in order to store, disseminate, absorb, transfer and refine useful information and knowledge across the firm.
- Organizational structure. Refers to the formal organizational design, and includes formal mechanisms for structuring the firm.

Relational capital

At the present time, relational capital remains underexplored – at least, in comparison to the manner and depth of the other types of intellectual capital discussed above. This may be because relational capital probably has the most complex and heterogeneous nature among all the types of intangible assets. Nevertheless, an additional effort must be made, because, as Acedo *et al.* (2006) remarked, one of the most fruitful developments of the RBV will be the 'relational one'.

From the knowledge-based view of the firm (Kogut and Zander, 1993), it has been highlighted that organizations are social entities which store internal and external knowledge, and this lies at the core of a firm's survival and success. When describing relational capital, we need to focus on how firms can absorb and exploit new knowledge from its environment in order to obtain and sustain a position of competitive advantage.

In this sense, before making comments about the concept of relational capital, its structure and components, we must review some of the key proposals to be found in the literature – inserted into the intellectual capital framework – in recent years.

One of the earliest and most important efforts in the identification and measurement of relational capital was made by Kaplan and Norton (1992). They focused exclusively on customer relationships, but we think that their vision of relational capital is too narrow. While for most firms this agent has the most strategic value, we cannot ignore other external relationships when explaining competitive advantage.

A similar narrow focus is also found in the Skandia Navigator (Edvinsson and Malone, 1997), which focused on customer relationships. Nevertheless, this work offers a wider set of variables in its relational component, such as: (i) customer typology; (ii) customer loyalty and longevity of relationship; (iii) customer support; and (iv) efficiency of customer relations.

In parallel, Brooking (1996) developed an interesting study about external firm relationships beyond those with customers. Brooking identifies other relational assets such as product/service brands, corporate reputation and image, or business partnerships and alliances, as well as including a detailed study of the customer relationship.

Following these efforts, we would like to draw attention to the Spanish proposal coming from the Intellect Model (IU Euroforum Escorial, 1998), which identified, in the case of relational capital, different elements or variables relating to the following environmental agents: (i) customers: (ii) suppliers: (iii) allies; (iv) other social agents; and (iv) corporate reputation.

Finally, and following the previous effort, the Intellectus Model (CIC, 2003), split relational capital in two: relational business capital (which includes the analysis of organizational relationships with customers,

Table 2.5 Definitions of relational capital

Concepts	Authors
Included in structural capital, relational capital is the relationships developed with key customers	Edvinsson and Malone (1997)
Focused on customer capital: relationships with customers and suppliers	Stewart (1997)
Can be defined as the combination of knowledge incorporated in the organization and employees, as a consequence of the value derived from the relationships they maintain with market agents and with society in general	CIC (2003)
The relationships an organization has with its clients/customers and environment	Martínez-Torres (2006)
Includes those intangible assets the company obtains when it supports relations with agents of its environment, such as clients, suppliers or allies	Alama Salazar (2008)
The knowledge embedded in relationships with customers, suppliers, industry associations or any other stakeholder that influence the organization's life	Cabrita and Bontis (2008)
Represents the knowledge embedded in the relationships with the outside environment	Chang et al. (2008)
Refers to customer capital, which represents the potential an organization has as a result of ex-firm intangibles	Wu et al. (2008), based on Bontis (1998)
Includes the value of all stakeholder, customer and supplier relations	Hsu and Fang (2009)

suppliers, allies and competitors, as well as other business-related activities); and relational social capital (referring mainly to the value of the firm's relations with society in general).

In the same way, Bontis (1999) also expanded the concept of 'client or customer capital' to include all the external relationships of the firm (for example, suppliers, allies, trade unions and so on).

Table 2.6 includes a literature review of the different aspects of relational capital, as well as the main assets/variables included in the proposals.

Table 2.6 Aspects of relational capital

Authors	Label	Main aspects included
Kaplan and Norton, 1992	Customer perspective	Corporate reputation and image Quality of customer relationships Product/services attributes Market share Customer loyalty Customer satisfaction
Brooking, 1996	Market assets	Product brand Customers Corporate image Social name Marketplace Business partnerships
Edvinsson and Malone, 1997	Customer capital	Customer typologies Longevity of customer relationships Customers' role Customer support Customer efficiency
Sveiby, 1997	External structure	Customer segmentation Efficiency Market growth Stability
IU Euroforum Escorial, 1998	Relational capital	Customers Brand Corporate reputation Alliances Relationships with other agents
CIC, 2003	Relational business capital	Relationships with: Suppliers Shareholders, institutions and investors Allies Business competitors Quality institutions
	Relational social capital	Relationships with: Public administrations Mass media and corporate image Green agents Social agents Corporate reputation
Alama Salazar (2008)	Relational capital	Customer relationships Supplier relationships Allies Corporate reputation

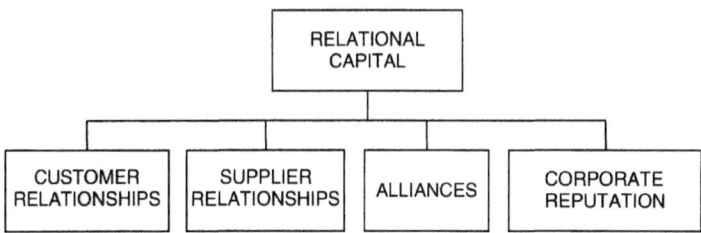

Figure 2.4 Components of relational capital

To point out the strategic value of relational capital, we highlight that it might be useful for the firmbecause: (i) it provides an external or market valuation of its existing knowledge base; and (ii) it provides information about market needs and opportunities, competitive dynamics and so on, thus supplying a useful external map or guide to ways that the firm can improve and develop new knowledge.

In this vein, relational capital could be critical for making decisions about how to exploit the current organizational knowledge base, and about detecting market trends and 'technological opportunity' (Kogut and Zander, 1992). Therefore relational capital can be considered a close relative of the well-known notion of 'dynamic capabilities' proposed by Teece *et al.* (1997).

At this time, and based on the literature review, we can note four basic components of relational capital (see Figure 2.4): (i) knowledge base from customers; (ii) knowledge base from suppliers; (iii) knowledge base from allies; and (iv) corporate reputation.

3
Technological Innovation

Technological innovation: concept and evolution

Since on the 1950s, scholars and practitioners around the world have expended much time and effort in defining and understanding the management of innovation. These efforts are seen in parallel with the growing importance of knowledge assets as key economic production factors.

The growing importance of knowledge as a productive factor in today's society requires a change in the thinking about innovation in general, and about specific terms, such as technological innovation, product innovation and organizational innovation (Nonaka, 1994). Company managers are aware that knowledge is the most valuable and strategic resource with which to face to current turbulent environment (Ordóñez, 2004; Chen *et al.*, 2004). These issues suggest that the rapid development from an industrial society to a knowledge society is defined by the importance of the creation of knowledge (Bueno *et al.*, 2004).

In addition, current global competition and the changing environment mean that innovation plays an important role, and companies must rapidly create new products, services and processes (Tushman and Nadler, 1986) to be competitive and to be able to adapt themselves to their environment (Damanpour, 1987). In this way, if a firm is to survive and thrive in dynamic environments, it must be continually renewed (Danneels, 2002).

Furthermore, stressing the importance of the dynamic capabilities view in this study, Gatignon *et al.* (2002), based on Teece and Pisano (1994) and Nelson (1995), believe that innovation and technical change are the essence of dynamic organizational capabilities. Therefore innovations can be understood as the key source in adapting to market change

(Stieglitz and Heine, 2007). In this sense, the success of the innovation process depends on a company's ability to exploit its resources, but above all the exploitation of previously non-existent dynamic capabilities, or their development (Verdú-Jover *et al.*, 2005).

Taking into account the ideas discussed earlier, it is obvious that the character of the environment is one of the contextual factors that influence innovation (Damanpour and Gopalakrishnan, 1998; Koberg *et al.*, 2003) and the more that is changing, the higher the capability of innovation must be to stay in the market.

Interestingly, when Swart (2006) reviewed the literature, he noted that innovation, among other concepts such as intangible assets, embedded tacit routines, core competencies and knowledge creation, is an important consideration in explaining a company's assets that is continually creating value beyond physical and financial resources.

Now that the current situation of firms and their interest in innovation have been considered, the term 'innovation' will be analysed.

Concept of innovation

In a similar way to the situation outlined in the previous chapter, where the different components of intellectual capital were defined, the concept of innovation is not unanimously agreed among the different scholars that appear in the literature (see Table 3.1). If we use a simple and first approximation, the term 'innovation' can be understood as the action of introducing something new to the market.

Because of the great quantity of definitions that exist in the literature, it is necessary to group them according to some criterion and thus be able to analyse them. This process creates two groups: on the one hand there will be borne in mind those definitions that refer to innovation as a process or flow; and on the other hand, those that see innovation as a final result or stock (see Table 3.2). Later, we shall refer to those definitions that demand marketing to consider innovation as such.

After examining the different definitions of innovation, a possible conclusion is that the majority of the authors observe innovation as a process. These authors show innovation as an operation that is progressing through time, from a new idea, and finally comes to a concrete result. Therefore, when innovation is understood as a process, the final result is included. Nevertheless, the proposals that consider the definition of innovation as a final result (apart from Tödtling *et al.*, 2009) do not consider the way that the above-mentioned result is achieved.

Table 3.1 Innovation concepts

Authors	Concept
Schumpeter (1912)	Implementation of new combinations of materials and forces, which may include the introduction of a new and untested product or production method, developing a new market, the conquest of a new source of supply, and creating a new organization in any industry
Thompson (1965)	Generation, acceptance and implementation of new ideas, processes, products and services
Knight (1967)	Adoption of a new and significant change by an organization
Nelson (1968)	Process by which new products and techniques are introduced into the economic system
Myers and Marquis (1969)	Commercialization of an invention
Zaltman *et al.* (1973)	Any idea, practice or material appliance perceived as being new to the organizational unit that adopts it
Rowe and Boise (1974)	Successful use of products or processes that are new to the organization and that achieve results according to decisions made within it
Gee (1981)	Process by which, from an idea, an invention or the identification of a need, a product, technology or service is developed that is accepted commercially
Damanpour and Evan (1984)	Deployment of an idea related to a device, system, process, policy, programme or service that is new to the organization at the time of adoption
Tushman and Nadler (1986)	Creation of any product, service or process that is new to a business unit
Van de Ven (1986)	A new idea, which may be a combination of old ideas, a project that challenges the present situation, a formula or a unique approach that is perceived as being new by the individuals involved
Deward and Dutton (1986)	Idea, practice or device received by the relevant unit of adoption as new material
Damanpour (1991)	Generation, development and implementation of new ideas or behaviours
OECD (1992)	Transforming an idea into a successful market product, new or improved, or a business process in industry and trade, or a method of social service

Nonaka (1994)	Process in which the organization defines and creates problems, and develops new knowledge to solve them
EC (1995)	Means to produce, assimilate and operate a successful innovation in the economic and social fields that brings new solutions to problems and thus meets the needs of both individuals and society
Morcillo (1995)	Seeing what everyone sees, reading what everyone reads, hearing what everyone hears, to innovate is to make what no one has yet imagined
Noria and Gulati (1996)	Any policy, structure, method or process, or any product or market opportunity, seen by the manager of a unit as innovative
Freeman and Soete (1997)	Attempts to commercialize an invention (an invention being the discovery of new methods or materials, namely the discovery of new knowledge)
Escorsa and Valls (1997)	A new idea comes to life or is implemented
Galunic and Rodan (1998)	Re-conceptualization of an existing system to use resources, which is developed as a new way of generating income and potential
Damanpour and Gopalakrishnan (1998)	Adoption of a new idea or behaviour in an organization
Williams (1999)	Implementation of discoveries and inventions, as well as the process by which new results are born, whether products, systems or processes
Camelo *et al.* (2000)	Creation or acquisition of an idea or knowledge, and its introduction into the organization, which may materialize in the form of a new product, process or method
Nieto (2001)	The first application of an invention, which takes place when the first commercial transaction is carried out with the new products, processes or services derived from it
Johnson *et al.* (2002)	Successful introduction of an invention in production or the market
Chen *et al.* (2004)	Introduction of a new combination of essential factors of production in the production system. It involves new products, new technologies, new markets, new materials and new combinations

(Continued)

49

Table 3.1 Continued

Authors	Concept
Edvinsson *et al.* (2004)	Re-using existing knowledge and perspectives in combination with new knowledge, as an invention, and then commercializing and capitalizing on it
Carson *et al.* (2004)	Creation of new knowledge that has potential for practical application in the development of new products or processes
Subramaniam and Youndt (2005)	Identifying and exploiting opportunities to create new products, services or work practices
OECD (2006)	Introduction of a new or significantly improved product (or service) process, marketing method or methods in organizational practices within the company, in the workplace or in foreign affairs
Grant (2006: 408)	Initial marketing of an invention to produce and market a new product or service or use a new method of production
Adams *et al.* (2006)	The successful exploitation of new ideas
Birkinshaw *et al.* (2008)	The invention and implementation of a management practice, process, structure or technique that is new to the state of the art and is intended to further organizational goals
Hidalgo and Albors (2008)	A problem-solving process, involving relationships between firms and different actors. An interactive, diversified learning process
Zheng (2008)	Encompasses creativity, knowledge creation, innovativeness, and innovation generation and implementation
Egbetokun *et al.* (2009)	A process by which firms master and implement the design and production of goods and services that are new to them, irrespective of whether they are new to their competitors, their countries or the world
Tödtling *et al.* (2009)	The result of an interactive process of knowledge generation, diffusion and application

Source: Based on Sousa (2006).

Table 3.2 Innovation concept typology

Innovation as a process	Schumpeter (1912); Thompson (1965); Knight (1967); Nelson (1968); Myers and Marquis (1969); Rowe and Boise (1974); Gee (1981); Damanpour and Evan (1984); Tushman and Nadler (1986); Damanpour (1991); OECD (1992); EC (1995); Morcillo (1995); Freeman and Soete (1997); Galunic and Rodan (1998); Damanpour and Gopalakrishnan (1998); Williams (1999); Camelo *et al.* (2000); Nieto (2001); Johnson *et al.* (2002); Chen *et al.* (2004); Edvinsson *et al.* (2004); Subramaniam and Youndt (2005); OECD (2006); Grant (2006); Birkinshaw *et al.* (2008); Hidalgo and Albors (2008); Zheng (2008); Egbetokun *et al.* (2009)
Innovation as a result	Zaltman *et al.* (1973); Van de Ven (1986); Deward and Dutton (1986); Noria and Gulati (1996); Escorsa and Valls (1997); Adams *et al.* (2006); Tödtling *et al.* (2009)

In this sense, Boer and During (2001) believe that innovation is a process and that it can be managed as such, so that the aim, design and organization of innovation should be formulated at the beginning, and its progress supervised.

On the other hand, some proposals contain certain peculiarities. In the case of Van de Ven (1986), he referred to innovation from a different point of view from the one presented in the Figure 3.1. He observed it as a process and defined it as the development and introduction of new ideas by means of people who, over time, take part in transactions with others within an institutional context. Furthermore, in Van de Ven's work, he uses the terms innovation and innovation process interchangeably, though at one point he defines innovation as a result.

Nevertheless, Johnson *et al.* (2002) were more specific and, in spite of observing innovation as a process, make a distinction between innovation and the innovation process. The latter term consists of the activities and results that take place between the conception of an idea and its introduction on the market, though if it is looked at in a more general way, diffusion and imitation of an innovation are included.

Tushman and Nadler (1986) went more deeply into the concept of innovation and suggested that innovation is complex and uncertain, changing through time and needing close collaboration, not only with R&D, but also with marketing, sales and operations activities.

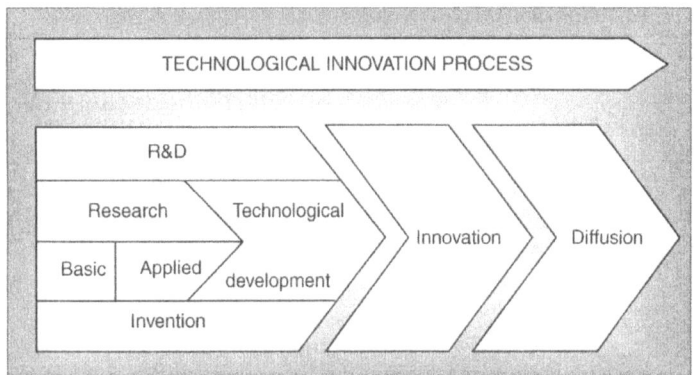

Figure 3.1   Technological innovation process
Source: Nieto (2001: 61).

Because this research has employed the intellectual capital-based view and, as it has been commented on several times already, intellectual capital is considered to be a different manifestation of organizational knowledge, it could be interesting to emphasize those studies in which the innovation process is referred to in an explicit way as knowledge management.

In several pieces of research, innovation is shown to be a knowledge management process (Camelo *et al.*, 2000; Carson *et al.*, 2004; Edvinsson *et al.*, 2004; Subramanian and Youndt, 2005). Furthermore, Nonaka (1994) asserted that innovation is the key in the creation of new organizational knowledge. In the same way, Nahapiet and Ghoshal (1998) proposed two main ways of knowledge creation: (i) through an incremental change and the development of existing knowledge; and (ii) with a radical change, through innovation.

Nevertheless, Nieto (2001) presented a narrow definition of innovation, and while he refers to the production of new knowledge, he highlights that innovation strictly occurs when the first commercial transaction takes place.

Referring to the last point – commercialization – several proposals, in an explicit way, such as, for example, Myers and Marquis (1969), Gee (1981), OECD (1992), Freeman and Soete (1997), Edvinsson *et al.* (2004), OECD (2006) and Grant (2006); as well as in an implicit way, as Adams *et al.* (2006), considered the necessity of taking into account commercial transactions in defining the concept of innovation, since the value of innovation depends on market success. In a deeper analysis of this issue, Jiménez and Sanz (2006) remarked on the distinction between product innovation – innovation occurring at the moment of its commercialization – and process innovation – innovation occurring at the time of the innovation's first industrial application.

Then, and taking into account previous literature proposals, as well as the intellectual capital framework, we could define *innovation* as the result by means of which, being fundamentally an intellectual capital-based process, new products and/or processes are created and/or substantially improved.

Finally, we would like to point out that our research is focused on technological innovation, which refers to changes made in the processes and products of a firm. The technology definition adopted in this research, based mainly on Navas (1994), Morcillo (1997) and Nieto (2001), uses in its strategic terminology resources, capabilities, knowledge, methods and so on that embody the concept of technology (see Table 4.1).

On the other hand, it is important to distinguish between generation and diffusion processes of new technologies – the innovation processes, and knowledge stocks a firm can possess – its technological endowment. So, because the process of change refers to technology levels, innovation is usually called technological innovation. Furthermore, and as several authors propose, administrative innovation is also considered (see the next section). Nevertheless, in the proposal made by OECD (2006), the term 'technological' has been deleted from the conceptualization of innovation because it is possibly a wrong interpretation as 'materials and high technology equipment handling'.

Finally, Zheng (2008) remarked on the importance of knowledge creation in innovation, because any new process, product or service derives from new ideas.

Innovation process

The view of innovation as an interactive process is nowadays broadly accepted among both scholars and practitioners (Tödtling et al., 2009). Different approaches (innovation networks, the innovative milieu approach, national/regional systems, clusters and so on) share this interactive view of innovation.

We can see innovation as a continuous interactive process, because of the importance of the study of innovation networks, since they are more explicit on the kinds of knowledge sources and types of interactions and links involved in that process.

Based on the distinction between formal/informal relations, and static/dynamic knowledge interactions, Tödtling *et al.* (2009) found four types of knowledge interaction in the innovation process:

- Market relations. This refers to the buying of 'embodied' technology and knowledge in various forms, such as the buying of machinery, information and communications technology (ICT) equipment or software, or licences.
- Knowledge externalities and spillovers. Different from market links, there is no contract or formal compensation for the acquired knowledge.
- Compared to market links, networks are more durable and interactive relations between specific partners in the innovation process. This constitutes a dynamic process of collective learning, where a given technology or piece of knowledge is not only exchanged but collectively further developed and the respective increased knowledge base.
- Informal links among firms and other types of organization, such as those in industrial districts and high-tech regions – public or private research institutions, for example. Such relations are particularly based on trust, shared understanding of common rules and behavioural norms. The include social capital.

As has been noted previously, in the scientific literature we can find many definitions of technological innovation. In a wide sense, Navas (1994: 38) defined it as 'the application of technology to different firm's goals'. Nieto (2001: 59) saw it as a process – as Figure 3.1 above – where inputs are knowledge-intensive, and the output is the new technological knowledge. In Nieto's words, 'technological innovation is associated with the idea of techno-scientific knowledge flow – generation, application, and diffusion'.

Table 3.3   Types of knowledge interaction in the innovation process

	Static (knowledge transfer)	Dynamic (collective learning)
Formal/traded relation	Market relations (relational capital)	Co-operation/formal networks (relational capital)
Informal/untraded relation	Knowledge externalities and spillovers (reputation)	Milieu/informal networks (social capital)

Source: Based on Tödtling *et al.* (2009).

Nieto (2001) and Hill and Rothaermel (2003) considered invention to be the first knowledge result arising from technology or applied science, and suppose a systematic of scientific knowledge, as the case of research and development (R&D). Nevertheless, it is important to highlight that R&D activities represents only a part of the innovation process, since firms can innovate through other learning mechanisms not necessary related to R&D.

So, the technological innovation process can be identified as a set of activities related to the production of innovations; that is, with R&D activities, but taking into account that the latter represent only a partial view of the process.

However, to innovate successfully, we must bear in mind that in addition to other competencies it is essential to possess the capability of R&D as a preliminary step. This idea coincides with the process of technological innovation introduced by Nieto (2001), where it can be seen that research and technological development precedes innovation.

However, on the other hand, to be capable of innovation, Dutta *et al.* (2005) interpreted the ability concept as the efficiency with which a company uses its available inputs and converts them into a desired output, presenting R&D expenses as an example of input (as a measurement of R&D capability) and the development of technological innovation as an example of output (as a measurement of innovation). Harris (2001) and Johnson *et al.* (2002) also referred to the inputs and outputs of knowledge, stating that many studies analyse the relationship between inputs of knowledge, using R&D expenses as a proxy, and outputs, using patents, the success of innovations or the quantity of new products as a proxy.

Continuing earlier arguments, it is interesting to note that Amit and Schoemaker (1993) considered R&D capability to be a strategic asset, which adds importance to the argument that has been presented, because our research can be linked with the study of intangible resources and capabilities that can play an strategic role in companies.

On the other hand, Johnson *et al.* (2002) introduced two points of view about the innovation process: (i) regarding its narrow definition, the process is outlined in the activities and results that take place between the conception of an idea and its market introduction; and (ii) with respect to its broad definition, they include both the dissemination and the replication of an innovation. Similarly, Grant (2006) also believes that the innovation process starts with a basic knowledge that leads to an invention by the development of new knowledge, and if that invention is produced and marketed (a new product or service) or used (a new production method), we could labelled it as an innovation.

Then the process of technological innovation can be understood as an increase in knowledge or a learning process as a part of a knowledge base or endowment to achieve a more developed, refined or advanced innovation. That is, the essence of the process of technological innovation is the accumulation of knowledge over time (Nieto, 2001).

This research supports that idea, studying how the different components of intellectual capital (as an input of knowledge) lead to a firm's innovation (output), considering such a course, as a whole, as a process of technological innovation.

The process is defined as a set of activities that help to produce new products and services, or to initiate new ways of production, including activities such as R&D and all creative tasks being carried out in a systematic way with the aim of increasing the volume of technological knowledge (Nieto, 2001).

Innovation typologies

Having discussed the concept of innovation and the innovation process, the next step is to distinguish different types of innovation, since there are several criteria for its classification. These criteria are shown in Table 3.4, along with their authors. These classifications of innovation will be analysed below.

The differentiation between innovation types is essential for a firm, because varying types of innovation play different roles in the firm's product portfolio and require different management approaches (Hurmelinna-Laukkanen *et al.*, 2008).

Technological and administrative/management innovations

According to Damanpour (1987), and Damanpour and Gopalakrishnan (1998), technological innovations are related to changes in products, services and production process technologies; that is, they are related to the primary activities of the firm, being possible in both products and processes. However, administrative innovations require changes in organizational structures and administrative processes; that is, they are related indirectly to primary and direct activities connected to the management of the company. Therefore, technical innovations are related to new technologies, products and services, whereas administrative innovations are linked to new administrative procedures, policies and organizational forms (Van de Ven, 1986).

Table 3.4 Innovation typologies

Criterion	Innovation typology	Authors
Based on the nature of innovation	Technological/ administrative or management innovations	Zmud (1984); Van de Ven (1986); Damanpour (1987); Navas (1994); EC (1995); Morcillo (1997); Damanpour and Gopalakrishnan (1998); Nieto (2001); Hill and Rothaermel (2003); Galende (2006); OECD (2006); Stieglitz and Heine (2007); Birkinshaw *et al.* (2008)
Based on the final results of innovation	Product/process innovations	EC (1995); Morcillo (1997); Tidd (2001); Nieto (2001); Boer and During (2001); Damanpour and Gopalakrishnan (2001); Adner (2002); Danneels (2002); Wang and Ahmed (2004); OECD (2006); Alegre and Chiva (2008)
Based on the degree of originality or disruptiveness	Radical/incremental innovations	Deward and Dutton (1986); Tushman and Nadler (1986); Henderson and Clark (1990); Morcillo (1997); Damanpour and Gopalakrishnan (1998); Chandy and Tellis (1998); Tidd (2001); Nieto (2001); Boer and During (2001); Gatignon *et al.* (2002); Darroch and McNaughton (2002); Koberg *et al.* (2003); Hill and Rothaermel (2003); Subramanian and Youndt (2005); Miller (2006); Laursen and Salter (2006); Stieglitz and Heine (2007); Hurmelinna-Laukkanen *et al.* (2008)
	Modular/ architectural innovations	Henderson and Clark (1990); Galunic and Eisenhardt (2001); McEvily and Chakravarthy (2002); Gatignon *et al.* (2002); Stieglitz and Heine (2007)
Based on its origin	Pull/push innovations	Zmud (1984); Cooper (1985); Morcillo (1997); Li and Calantone (1998); Darroch and McNaughton (2002)
Based on the effects of innovation on firms' competences	Competence-enhancing/ competence-destroying innovations	Schumpeter (1942); Anderson and Tushman (1990); Gatignon *et al.* (2002); Rothaermel and Hill (2005)

These innovations are based on different processes. Whereas technological innovation originates in the technical core and follows a bottom-up process, administrative innovation stems from the administrative core and remains a top-down process (Damanpour and Gopalakrishnan, 1998). Moreover, it is important to distinguish the two types, since they have different aims, procedures, activities, technologies and are affected by different forces in the environment (Zmud, 1984).

On the other hand, the organization's general receptiveness toward the change exerts a greater influence on the process of technical innovation than the one on the process of administrative innovation, because the latter often influences the behaviour of company managers while technical innovation influences behaviour throughout the organization (Zmud, 1984).

Morcillo (1997) and OECD (2006) referred to innovation in management methods and organizational innovation, respectively. Both types can be equated to administrative innovation, since the first is defined as innovation developed in commercial, financial and organizational areas, and the second relates to the introduction of new organizational practices, the layout or external affairs of the firm.

However, while Morcillo includes commercial aspects within this type, the OECD's Oslo Manual considers another kind of innovation. This is labelled 'marketing innovation', which is defined as the application of new marketing methods involving significant changes to the design or packaging of a product, its positioning, promotion or pricing.

Furthermore, EC (1995) noted changes in management, job organization, working conditions and workers' skills, which are implicitly referred to as administrative innovation.

Birkinshaw *et al.* (2008: 826) defined management innovation as 'a difference in the form, quality, or state over time of the management activities in an organization, where the change is a novel or unprecedented departure from the past'.

Modular and architectural innovations

Henderson and Clark (1990) propose a classification that examines incremental and radical innovations in depth, since they distinguish between the components of a product and how these are integrated into the system – what they call 'product architecture'. Therefore, an architectural innovation is one that changes the way of linking the various components of a product, so its essence is the reconfiguration of an established system, to link components in different ways. Thus, this

Table 3.5 Specific innovation framework

		USED COMPONENTS	
		Reinforced (Current)	Overturned (New)
COMPONENTS ARCHITECTURE	**Without changes**	Incremental innovation	Modular innovation
	With changes	Architectural innovation	Radical innovation

Source: Based on Henderson and Clark (1990: 12).

type of innovation is a challenge for established firms and may have important competitive implications.

Because that classification is closely related to radical and incremental innovation, Henderson and Clark presented a specific framework for the definition of innovation (see Table 3.5). Similarly, McEvily and Chakravarthy (2002), based on Christensen (1992), presented a matrix where the types of innovation are the same as in the table, highlighting that incremental innovation is a functional knowledge modification; architectural innovation represents a functional knowledge recombination; modular innovation is a change at the component level; and radical innovation is a change at the product level. In addition, the first two are included within 'the level of knowledge of the existing component' and the latter two under 'the new knowledge of acquired component'.

In this sense, Stieglitz and Heine (2007) developed an adaptation of the previous model, highlighting technological resources rather than talking in general terms. They also suggest greater attention should be paid to modular innovations, stating that they fundamentally change the technology architecture of established products.

Gatignon *et al.* (2002) also referred to this type, though using a different nomenclature. These authors studied changes in subsystems and/or mechanisms associated with products, stating that architectural innovation involves changes in the existing links among subsystems, and generational innovation (modular for Henderson and Clark, 1990) involves changes in subsystems that are related to the existing involved mechanisms.

In addition, they structure the various aspects related to innovation, because they think there is confusion about the nature, types and hierarchy of innovation, highlighting that both architectural and generational innovation are types of innovation.

Therefore, unlike the two previously-quoted works, Gatignon *et al.* (2002) did not study that classification of innovation together with radical and incremental innovation, because while the former refers to types of innovation, the latter classification refers to the intrinsic characteristics of innovation.

Finally, we draw attention to the fact that Galunic and Eisenhardt (2001) believe that dynamic capabilities reconfigure the division of resources; that is, make possible architectural innovation. However, these authors provide a different view of the phenomenon, since they refer to organizational forms rather than products, thus relating architectural to administrative innovation.

Radical and incremental innovations

The distinction between radical and incremental innovation is a widespread typology in the literature. However, Henderson and Clark (1990) think it is potentially misleading and incomplete, and it does not explain some important effects that occur in firms and industries derived from minor improvements in products.

These authors believe that incremental innovation introduces relatively minor changes with respect to existing products, exploits established design potential and usually strengthens the dominance of established firms; and that radical innovation is based on different engineering systems and scientific principles, and often opens up new markets and potential applications. The two types of innovation have different competitive effects because they require different organizational capabilities.

Generally, radical innovation is based on new concepts of design that break with existing paradigms, whereas incremental innovation is based on minor changes or improvements to the present technology (Li *et al.*, 2008).

According to Dewar and Dutton (1986), Nieto (2001) and Tidd (2001), this classification is based on the degree of originality and novelty of the innovation process. Radical innovation occurs when new products/ processes that are introduced are completely different from existing ones, whereas incremental innovation occurs through small incremental improvements in products/processes. Therefore, the first type is given the opportunity to accumulated scientific and technological knowledge, and the second represents continuity with existing technologies. On the other hand, Koberg *et al.* (2003), in defining such innovations, took into account not only products and services, but also markets. Finally, in addition to products, processes and markets,

Miller (2006) also referred to business models and industry structures in the definitions of both types of innovation.

Hurmelinna-Laukkanen *et al.* (2008) believe there are some common elements related to radical innovations. The definitions typically include aspects related to high market and technological uncertainty, new market creation, product cannibalization, and even effects on the organization's knowledge repositories.

Based on Dewar and Dutton (1986), Song and Thieme (2009) studied in depth the concept of radical innovation, proposing that it represents revolutionary changes in technology: "These innovations incorporate technology that is a clear and risky departure from the state of current knowledge prior to the introduction and have a high degree of new knowledge embodied in the technology" (p. 49). Those revolutionary changes in the technology mean new technology that has the potential to achieve one or more of the following: (i) a completely new set of the resulting characteristics; (ii) five or more improvements in well-known characteristics; and (iii) a significant reduction (30 per cent or more) in costs.

In relation to this classification, Subramaniam and Youndt (2005) referred to the concept of innovative capabilities, distinguishing between incremental and radical innovative capabilities. The first is defined as the capability to generate innovations that refine and reinforce existing products and services, while the second is the capability to generate innovations that significantly transform existing products and services.

In addition, Tidd (2001: 177) related what he called type of innovation (process, product, service) with the degree of innovation (disruptive, radical, incremental), creating a matrix using two variables, labelled 'area of innovation'. At the same time, Nieto (2001) and Morcillo (1997), within the definition of product and process innovation, highlighted radical and incremental innovation, which suggests that there is an interesting link between the two forms of classification.

Then, by linking the two variables, nine different types of innovation can be perceived. Booz *et al.* (1982) distinguished the following categories: (i) improvements in existing products; (ii) new product lines; (iii) expansion of existing product lines; (iv) new global products; (v) reduction in costs by process development; and (vi) the development of the product portfolio (repositioning). This differentiation also represents both incremental and radical innovations, such as product and process innovation, spanning a range from simple improvements to existing processes and products to radical innovations.

The peculiarity offered by Tidd (2001) was that, regarding the kind of innovation he gives a separate treatment to services, and regarding the degree of innovation adds a new level called 'disruptive'. This author, referring to Christensen (1997), defined disruptive innovation as one that provides a different system of features that will probably appear in very different segments within the market. In the same sense, Govindarajan and Kopalle (2006) showed that disruptive innovation introduces a different system of features and performance attributes with respect to existing products, offering them at a lower price, because the combination is not attractive to established customers at the time the product is introduced, as a result of poor performance in the attributes that add value to these customers. Despite this, a new segment of customers (or the established market that is more sensitive to price) perceives value in the new attributes of innovation and the low price, but as time passes, later developments increase the attributes of the new product to a sufficient level to satisfy established customers.

Nevertheless, taking into account Henderson and Clark's (1990) explanation about radical innovation, by means of which new markets emerge, it can absorb a disruptive innovation. Tushman and Nadler (1986) also related, in a similar way to Tidd (2001), the degree of innovation to innovation (see Table 3.6).

Regarding the levels of innovation that Tushman and Nadler (1986) and Boer and During (2001) propose, it is necessary to clarify that, when talking about product innovation, 'synthetic' relates to what Henderson and Clark (1990) called architectural innovation (see Table 3.5, above)

Table 3.6 Innovation typology matrix

	Product	**Process**
Incremental	Provides additional features, new versions or extensions to what would otherwise be a standard product line	Improvements achieved with lower costs, better quality or both
Synthetic	Refers to the combination of existing ideas or technologies used creatively to develop new products	Refers to significant increases in the size, volume or capacity of production processes that are well-known
Discontinuous	Involves the development or application of new and significant ideas or technologies	Totally different way of producing products or services

Source: Based on Tushman and Nadler (1986).

described as discontinuous innovation, of both product and process, it refers to radical innovation, explained above.

Chandy and Tellis (1998: 476) developed a matrix in which the dimensions 'technology' and 'market' were considered. They consider that incremental innovation involves relatively minor changes in technology and provides relatively low incremental benefits to customers, while radical innovation involves new technology and provides greater benefits to customers compared to existing products.

Damanpour and Gopalakrishnan (1998) defined incremental and radical innovation based on technical and administrative innovations rather than product and process innovations. Radical innovation produces changes in organizational activities and represents a clear divergence from existing practices, while incremental innovation involves a lesser degree of deviation. However, Hill and Rothaermel (2003) defined them from a technological point of view.

On the other hand, Dewar and Dutton (1986) think that radical and incremental innovation are different types of technological innovation process, and the major difference between them is the degree of novelty of the technological process embedded in innovation – and thus the degree of knowledge embedded in innovation. Similarly, Gatignon *et al.* (2002) referred to two types involving a technical background, though they do not see them as types of innovation, but rather as characteristics of innovation.

Furthermore, Laursen and Salter (2006) suggested that often a firm needs to make a significant investment in R&D to achieve a radical innovation, involving a lower chance of success but higher performance, whereas an incremental innovation is more common but involves lower performance.

Finally, it is interesting to explain the different meanings that, according to Nieto (2001), have incremental and radical terms (see Table 3.7.)

The magnitude of the impact is related to the impact of innovations in the socio-economic environment, so that radical innovations noticeably alter the structure of the emerging sectors, while incremental ones do not cause major upheavals in the industry in which they appear,

Table 3.7 Meaning of incremental and radical innovation

Type of innovation	Degree of impact	Nature of the process
Incremental	Low	Continuous
Radical	High	Discontinuous

Source: Nieto (2001: 94).

nor do they alter the competitive position of established companies. For its part, the nature of the innovation process explains how to generate innovations: how to increase the technological knowledge flow, transforming it into new products and processes.

In this sense, it is essential that companies distinguish between radical and incremental innovation, since they have different roles in the product portfolio of the company and require different management approaches. While the most radical innovations have the potential to transform whole industries, incremental innovations provide a low-risk potential in order to improve the category of products (Hurmelinna-Laukkanen *et al.*, 2008).

After submitting that there are different ideas about this type of innovation, we can see that radical and incremental innovation are at opposite ends of a line that represents the novelty of innovation. Incremental innovation would appear on the far left and radical innovation would appear on the far right of the line; that is, the degree of novelty would be represented from less to more.

Market pull and technology push innovations

When innovations have their origin in market orientation, they are called market pull, and when science and technology generate innovations, these innovations are called technology push (Morcillo, 1997). The first type start from the analysis of customer needs, and the second result from the dynamics of technological research (Verganti, 2008).

Specifically, Li and Calantone (1998) asserted that the company should gain knowledge of the market through existing customers. The authors argued that competences about market knowledge, as a process that generates and integrates such knowledge, have a positive influence on the advantage derived from new products.

Moreover, several authors have studied this typology with reference to other criteria. Darroch and McNaughton (2002) stated that incremental innovations are usually classified as market pull innovations, because of emerging market needs, while radical innovations start from science or originate in the company, and are classified as technology push innovations. Furthermore, they argued that market pull innovations usually flow from companies that are market-orientated.

In addition, when Verganti (2008) represented innovation strategies, he overlapped market pull with incremental improvements and technology push with radical improvements.

In the same sense, Morcillo (1997) used product and process innovation to explain what is meant by market pull and technology push innovation, respectively. While product innovation results solely from a market orientation by the company attempting to adapt to the potential demand development, the innovation process responds to an internal orientation that has a priority focus on cost. Finally, Zmud (1984) argued that market pull innovations have a higher chance of commercial success than technology push ones.

On the other hand, Darroch and McNaughton (2002) suggested that technology push innovations usually threaten the business, because it is more difficult to achieve commercial success with these, therefore generally market pull innovations have a greater chance of success from the commercial point of view (Zmud, 1984). However, technology push innovations are more important in order to achieve success over a longer period of time, since they can change the existing market structure.

Finally, Zmud (1984) and Darroch and McNaughton (2002) discussed the need to consider both types of innovation complementarily. In this way, they posit that a higher level of innovations will occur when needs identified in the market (pull) and the means to address such needs (push) occur simultaneously at this point.

Therefore, companies that use a well-balanced mix of both types of innovation, will have better results than those that focus solely on market pull or technology push innovation (Cooper, 1985).

Competence-enhancing and competence-destroying innovations

When Gatignon *et al.* (2002) examined a structural approach to assessing innovation, as well as considering radical and incremental innovation, they proposed another point of view, which distinguishes between competence-enhancing and competence-destroying innovation.

In this sense, Schumpeter (1942: 83) argued that new technologies create new market opportunities but at the same time damage or destroy demand in many existing markets. This is described as the process of creative destruction, a crucial aspect of capitalism.

Competence-enhancing innovation builds on and reinforces existing competencies, skills and know-how, and competence-destroying innovation supersedes and overturns existing competencies, skills and know-how (Gatignon *et al.*, 2002: 1107).

Also, Rothaermel and Hill (2005) suggested that competence-destroying innovations require new skills, capabilities and knowledge in the development and production of products and services, while competence-enhancing innovations are developed by firms' existing know-how and they do not supersede the skills required to manage the previous technology. In this regard, Anderson and Tushman (1990) think that competence-enhancing innovation introduces a new technical approach that leads to an increase in performance, and rather than supplant the existing approach, it builds on it.

Gatignon *et al.* (2002) think that these characteristics depend on the particular history of each company, since a given innovation could increase skills in some firms, but damage them in other companies.

This typology of innovation is closely related to incremental and radical innovation, since radical innovations can be critical, meaning that some companies come into an industry and can threaten established companies (Henderson and Clark, 1990).

Overview of innovation types and the choice of technological innovation

Once we have studied the criteria used to distinguish different types of innovations, it is interesting to reflect on these.

On the one hand, we shall examine the relationship that can exist between the various criteria used to classify innovations and, on the other hand, because this research aims to explore the causal relationship between intellectual capital and innovation, and because there are several kinds of innovation, it is necessary to make a choice regarding the type (or types) of innovation that this study will take into consideration.

Regarding the link that can occur between different types of innovation related to different criteria, some authors do not consider several criteria, but take into account several kinds of innovation when explaining one of them.

Darroch and McNaughton (2002) studied where innovation comes from, making reference to the degree of originality and novelty of the innovation process; that is, they link radical innovation and incremental innovation to technology push and market pull innovation, respectively. Morcillo (1997) used product innovation to explain market pull, and process innovation to explain technology push innovation. Finally, Henderson and Clark (1990) related radical innovation to competence-destroying innovation.

Some ideas that have appeared in the literature, and have been discussed in the text above, are presented in Figure 3.2.

Therefore, based on Figure 3.2, it can be asserted that (except for the criterion of the nature of the innovation process – technological/administrative) the different criteria on innovation are related to one or two criteria, and that these links lead to a better understanding of each type of innovation.

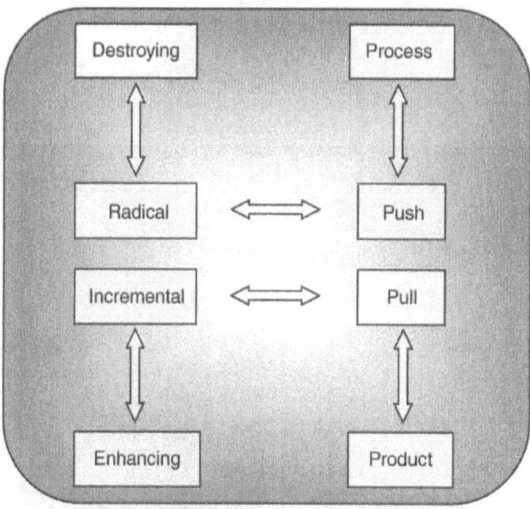

Figure 3.2 Assimilation among types of innovation

Regarding the choice of the type of innovation that will be considered in this study, the main reason is that technological innovation has been analysed, and according to Morcillo (1997) and Oslo Manual (2006), is the one that includes changes in products and processes. In addition, this typology is the most accepted and recognized in the literature. Specifically, product innovation is one of the most promising areas in the field of knowledge management (Corso *et al.*, 2001).

A final issue to be addressed, which is related to the joint study across several criteria, it is the analysis of the figure proposed by the Oslo Manual (OECD, 1997), and where technological innovations (product and process), administrative innovations, and changes that are not considered innovative because they do not have sufficient novelty level can be observed (see Figure 3.3).

So, like other authors (Tushman and Nadler, 1986; Tidd, 2001), the Oslo Manual (OECD, 1997) jointly studied different types of innovation (product/process/administrative and radical/incremental), adding various changes and improvements that cannot be considered innovative.

Moreover, this figure does not only refer to the type and degree of novelty regarding innovation, but also to the scope; that is, it takes into account whether the innovation has occurred worldwide, at the intermediate level (if it has been in the industry or geographical area

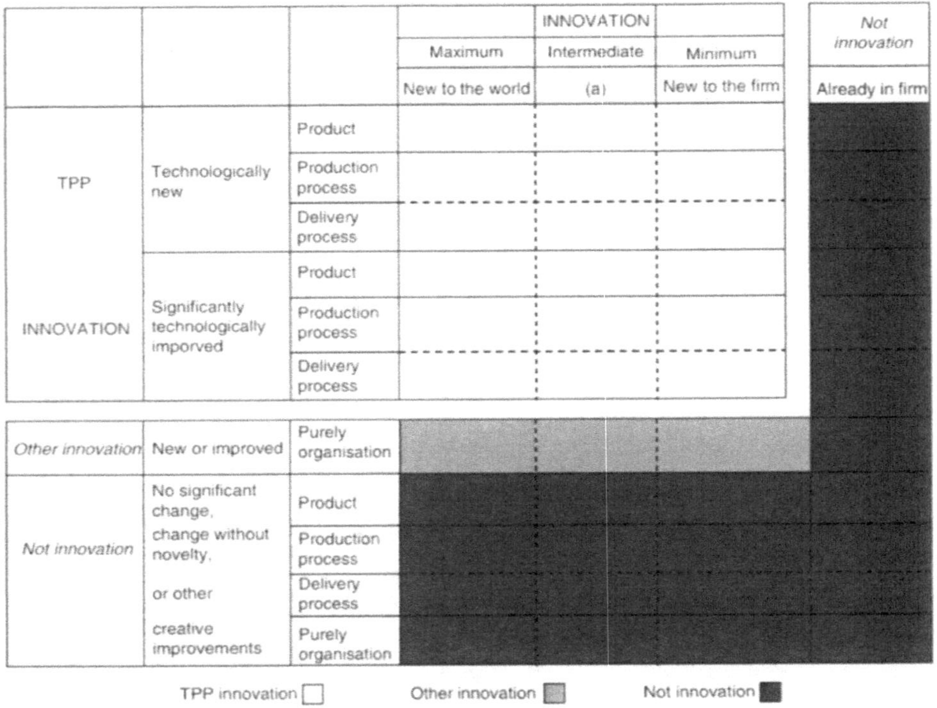

68

			INNOVATION			Not innovation
			Maximum	Intermediate	Minimum	Already in firm
			New to the world	(a)	New to the firm	
TPP INNOVATION	Technologically new	Product				
		Production process				
		Delivery process				
	Significantly technologically imporved	Product				
		Production process				
		Delivery process				
Other innovation	New or improved	Purely organisation				
Not innovation	No significant change, change without novelty, or other creative improvements	Product				
		Production process				
		Delivery process				
		Purely organisation				

☐ TPP innovation   ▨ Other innovation   ■ Not innovation

Figure 3.3 Type and degree of novelty and innovation definition
(a) Could be geographically, for example new to country or region
TPP: technological product or process innovation
Source: OECD (1997: 36).

where the firm acts) or at the level of the firm. In the same sense, Boer and During (2001) also considered for whom this is an innovation, but added three levels (new to the world; intermediate; and new to the firm) at the individual level.

Product and process innovation

This way of classifying types of innovation refers to the output of the technological innovation process. When new technological knowledge is embodied in new product development or the enhancement of existing states, we are talking about product innovation; and when it is embodied in the launch of new production processes and/or engineering changes in existing processes we called it process innovation (Nieto, 2001; Egbetokun *et al.*, 2009).

Tidd (2001) also considered this typology of innovation, but when he offered both definitions, he did not refer only to products but also took services into account.

For its part, when EC (1995) explained innovation, it referred to the renovation and expansion of the range of products and services and relevant markets, which can be understood as product innovation; and the renovation of production methods, supply and distribution, which can be equated to process innovation.

Wang and Ahmed (2004) did not use exactly the same terminology when they used the term 'product innovativeness', defining it as novelty and a significant sense of new products introduced into the market at the right time; and 'innovativeness process', which includes the introduction of new production methods, new management approaches and new technology that can be used to improve production and management processes. However, with regard to the latter comment, this study does not share the idea of management issues within the process of innovation, which is part of administrative innovation, as was discussed in the previous section. It is also interesting to note that the authors add that these innovative products offer significant business opportunities in terms of growth and expansion into new areas. In addition, it seems that this terminology intrinsically considers the radicality of innovations, so it would be between the classification according to result and the degree of originality of innovation, as in the work of García and Calantone (2002).

Moreover, Morcillo (1997) grouped innovation into three broad categories: technological innovation, social innovation and management methods innovation, and within the first of these he differentiated between products and process changes, which are defined in line with Nieto (2001). The only difference is that Morcillo indicated the impact of process innovation, commenting productivity and streamlining of manufacture will improve, and therefore also the cost structure.

For his part, Adner (2002) argued that the effects of product and process innovations are reflected in changes of product functional outcome and product cost, respectively.

OECD (2006) showed the four main types of innovation taken into account: product, process, marketing and organizational innovation, and we have considered the first two exclusively in this section. In this book, the two definitions are discussed in greater depth than in previous works.

A product innovation corresponds to the introduction of a new product or service, or one significantly improved in terms of features, or intended use. This definition includes the significant improvement of the technical characteristics of components and materials, integrated computer science, ease of use, or other functional characteristics (OECD, 2006). Similarly, Alegre and Chiva (2008) specified the different issues considered within this type of innovation, and they defined product innovation as the successful exploitation of new knowledge, implying two conditions: novelty and use. Therefore product innovation is a process that includes technical design, R&D, manufacturing, management and commercial activities involved in the marketing of a new (or improved) product.

The exact circumstances in which this kind of innovation occurs are specified, noting that significant improvements in existing products occur when materials, components or other characteristics that make up these products are changed to improve performance. It is also thought that product innovation occurs when the technical specifications are slightly changed and a new use is developed for a product (OECD, 2006).

In addition, OECD (2006) asserted that, as defended by Tidd (2001) and Tushman and Nadler (1986), the term 'product' includes both products and services.

A process innovation is also the introduction of a new, or significantly improved, production or distribution process. This implies significant changes in techniques, materials and/or software (OECD, 2006); detailing again specific cases in which this type of innovation occurs.

As with Morcillo (1997), the Oslo Manual (OECD, 2006) also referred to the impact of this type of innovation, stating that it can be designed to reduce the unit costs of production or distribution, improve quality, or produce or distribute new (or significantly improved) products.

However, unlike Morcillo, the Oslo Manual places a special emphasis on distribution.

It is interesting to note the observation the Oslo Manual presents on service innovation, showing that a key aspect of services is that the distinction between product and process is often unclear, since production and consumption occur simultaneously. That is, it is usually a continuous process, consisting of a series of gradual changes in products and processes. It is therefore more difficult to separate product and process innovation when we talk about services.

Finally, Danneels (2002: 1105) analysed in depth product innovation and developed a theory of product innovation that simultaneously considered both the exploitation and exploration of customer and technological competences. That is, first, a new product can draw on existing technological competences or require new ones, and, second, draw on customer competences that the firm already has, or require a new type of customer competences.

Then, using the four variables listed, Danneels created a matrix using different types of new products (see Figure 3.4).

In pure exploitation, a firm uses both existing technological and customer competences and in pure exploration, the new product is a tool to build new competences relating to both customers and technologies. Regarding two intermediate cases, leveraging technology (exploiting technology/exploring customers) implies appealing to additional customers through developing products based on an already achieved technological competence, whereas leveraging customer competence

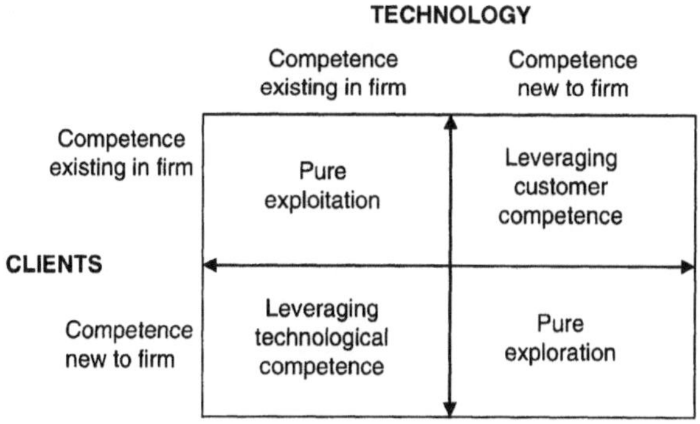

Figure 3.4 Competence-based new product typology
Source: Danneels (2002: 1105).

(exploiting customers/exploring technology) involves building additional technological competences to appeal to a greater share of existing customers' needs.

This typology can be interesting for the model presented in this study, since the technological and client competencies can be assimilated into technological and relational capital, respectively, and explore the links to product innovation.

4
The Role of Intellectual Capital
in Technological Innovation

Within the theoretical framework discussed, this chapter attempts to show the influence of the allocation of resources and capabilities, or, in our case, the elements of intellectual capital, in achieving a competitive advantage, understood as a result of the innovation process. Such a perspective of innovation based on intellectual capital also helps in gaining a better understanding of the innovation process, since, as noted by Galende and De la Fuente (2003), much of the research addressing the study of innovation from an external perspective does not consider the internal complexity that characterizes the process of innovation, and which allows them to analyse the intellectual capital. Moreover, we can consider that the ability of a firm to innovate depends very closely on the intellectual assets and knowledge owned, and how these can be deployed (Alegre and Lapiedra, 2005; Subramaniam and Youndt, 2005). Furthermore, though the initial links have been made between knowledge and/or intellectual capital and innovation, more research must be done in order to understand its complex and precise nature.

The specific objective of this research is to determine the influence of human capital (which includes aspects such as experience, skills, commitment and professional development of employees), structural capital (knowledge and technical skills of working groups and the organization, culture, values and structure), and relational capital (knowledge of the relationships developed with customers, suppliers, allies and so on) on product and process innovations. In this regard, having stated the theoretical framework, the working model, as reflected in Figure 4.1, includes three basic assumptions for later empirical contrast.

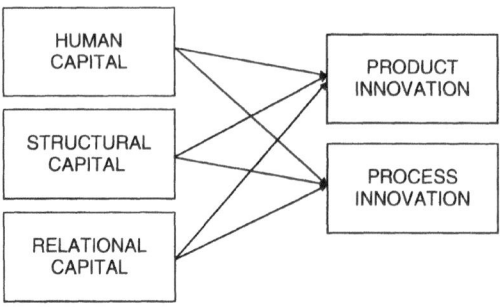

Figure 4.1   Causal model proposed

Intellectual capital and technological innovation

While it has been recognized that economic wealth comes from knowledge assets – intellectual capital – and its useful application replacing or perhaps supplementing land, labour and capital (Teece, 1998; Dean and Kretschmer, 2007), the emphasis on this idea is relatively new, and the management of a firm's intellectual capital has become one of the main tasks on the executive agenda. Nevertheless, this work is especially difficult, because of the problems involved in its identification, measurement and strategic assessment. In this situation, the models of intellectual capital become highly relevant, because they not only allow us to understand the nature of these assets but also to carry out their measurement.

Innovation is generally viewed as one of the most important sources of sustainable competitive advantage, because it leads to product improvements that increase the value of the product portfolio (Coombs and Bierly, 2006), helps firms to survive, makes continuous advances (Liu *et al.*, 2005), allows innovators to grow more quickly, be more (dynamically) efficient, and ultimately be more profitable than non-innovators (Mansury and Love, 2008). In fact, innovation has also been defined as the most knowledge-intensive organizational process, which depends on the individual members and the collective knowledge of the firm (Adamides and Karacapilidis, 2006).

While the basic link between intellectual capital and innovation in firms is on the whole persuasive, additional efforts to understand this causal relationship are worthwhile. We analysed innovation performance through new product/process innovation, because this becomes necessary for a firm's survival and sustained competitive advantage (Hsu and Fang, 2009).

In the following subsections we shall attempt to develop the main hypotheses regarding the influence of human, structural and relational capital on a firm's innovation.

The role of human capital in technological innovation

One of the key determinants of value creation for knowledge-intensive firms is the innovation created by their human capital (Edvinsson and Sullivan, 1996). Similarly, Wang and Chang (2005) asserted that human capital is the most important component of intellectual capital and knowledge along with the capabilities of employees as a source of innovation.

The review of the measurement of innovation management by Adams *et al.* (2006) refers to the importance of the human factor, taking into account the number of people assigned to innovation tasks relating to the propensity to innovate, skills, experience and education. In a similar vein, Amabile *et al.* (2007) argued that the responses of creative employees, who have original ideas for changing products, services and processes, are relevant to introducing innovations.

Having brilliant, motivated and experienced human capital should be the basis for the majority of innovation processes in an organization. Therefore this kind of intellectual capital should provide the main source from which to develop new ideas and knowledge (Snell and Dean, 1992; Pérez and Quevedo, 2006). Individuals and their associated human capital are crucial for exposing an organization to technology boundaries that increase its capability to absorb and deploy knowledge domains (Subramaniam and Youndt, 2005). Highly motivated and highly trained employees may question established organizational routines; hence, this kind of human capital becomes critical to push the firm to its technological limits, constituting the best incentive for obtaining new knowledge and innovation (Nonaka and Takeuchi, 1995; Hill and Rothaermel, 2003).

In the same sense, Wang *et al.* (2008) highlighted that human capital may be the key source of new ideas and knowledge within an organization, thereby increasing the likelihood of transforming prevailing knowledge. The level and specificity of human knowledge could either facilitate or limit the absorptive capacity of the organization. A critical portion of the know-how required for technological innovation resides with and is used by human capital.

A human resource management system to increase human capital stocks or endowments should rely on leadership and management ability, as well as on improving the attitude of the workforce, and increasing knowledge and skills (Tseng and Goo, 2005), and this, in turn, will have a positive effect on innovativeness. From this, it makes sense to state that those firms with the best human capital (the most trained, experienced and motivated) will be able to create the highest number of new ideas and products.

Moreover, some authors (for example, Baldridge and Burnham, 1975) studied demographic characteristics (such as gender, age and cosmopolitanism), concluding that these do not influence innovative behaviour; while other authors (for example, Amabile, 1998) show that innovative groups should be formed by individuals with several of these characteristics.

Some important research refers to Damanpour's (1991) meta-analysis, where different sources of innovation were studied, including worker professionalism. This element relates to the education and experience of employees, and was founded to have a positive and significantly statistical association with technical innovation.

On the other hand, with respect to process innovation, Damanpour (1991) also pointed out that professional knowledge, including the education and experience of employees, is related positively to this type of innovation.

Similarly, Hayton (2005) indicated a positive relationship between education and the functional diversity of employees and the 'innovation' dependent variable, but highlighting new processes. Thus, as in the previous arguments, Hayton argued that the diversity and heterogeneity of human capital represents a wide cognitive range which facilitates the acquisition of new knowledge. On the other hand, following the same logic as explained in the dependent variable, it presents the impossibility of finding a positive relationship between the senior executives' stock and new processes.

Regarding the professional experience included in 'qualified personnel recruitment', Díaz et al. (2006) found a positive influence on the likelihood of having innovative capabilities within the firm, including new processes, among other innovative aspects. Thus, as explained above, it represents the importance of knowledge owned by employees when they start to work in a new firm.

In his study on the effect of intellectual capital on the entrepreneurial characteristics of the firm, Hayton (2005) showed empirically the positive relationship between the diversity of human capital, both educational (considering areas of science, business and humanities) and functional (functional experience), and innovation; since diversity promotes the search for knowledge and organizational learning processes, and the different cognitive approaches promote creativity. This study also measured innovation, considering (among other aspects) the number of new products and services. However, it has not been shown that the stock of senior executives has a positive influence on innovation. This may be because it is not enough to have just a higher education and experience, it is also important to motivate employees to increase innovation.

Therefore, and in general terms, according to the arguments above, we expect human capital to enhance firm performance through innovation; therefore, the following subhypotheses are proposed.

H1a: *The higher the human capital endowments of the firm, the higher their product innovation.*

H1b: *The higher the human capital endowments of the firm, the higher their process innovation.*

The role of structural capital in technological innovation

Beyond human capital, an important part of the knowledge, abilities, experiences and behaviours required for the successful development of new products and processes lie within the organization. As Van de Ven (1986) pointed out, the innovation process – in general terms – is a collective achievement of the organization's members, where organizational support becomes a key element. Institutionalization acts as a means of preserving organizational knowledge and routines, which in turn fosters the accumulation, preservation and improvement of collective knowledge.

Following Tseng and Goo (2005), good structural capital will translate the human dimension of innovation into company property. To do so, firms must support and nurture the brightest individuals to share their innovation, knowledge and abilities through organizational learning. Tacit issues such as managerial commitment, a common identity and shared vision, or a climate of openness and experimentation, compose the learning capability of the firm (Akgün *et al.*, 2007). Nevertheless, while organized information cannot be a substitute for tacit knowledge, it can significantly enhance it to fill existing knowledge gaps; hence information technologies can support the innovation process (Adamides and Karacapilidis, 2006). In this vein, operational processes, information systems, organization culture, internal organizational structure, R&D efforts and administrative systems will have a positive influence on the innovative capabilities of the firm. Thus databases, procedure manuals, effective information systems, or cultural values devoted to innovation promotion can constitute important sources for innovative success.

Structural capital, on its 'organizational side' includes (Wu *et al.*, 2007) organizational processes which, among other things, have a positive effect on innovative performance, with reference to new products introduced by the firm in the market. Therefore, structural capital will be a source of product innovation if employees are encouraged and stimulated to create new ideas and to innovate to initiate new products, since it will increase their intrinsic motivation and enhance performance. In addition, within that explanation, the designed evaluation systems and organisational culture are considered, so the aspects of organisational strategy studied by Un and Cuervo-Cazurra (2004), who analysed organisational strategy, and the culture of innovation analysed by Prajogo and Ahmed (2006) and Pizarro *et al.* (2007), appear again.

Van de Ven (1986) analysed the innovation process and showed the importance of different factors. Within them he highlighted the implementation and institutionalization of innovative ideas as key determinants in the process. In the same sense, Salman and Saives (2005) suggested that the accumulation of knowledge enhances the organizational capacity to recognize and assimilate new ideas, as well as its ability to convert this knowledge into innovations.

Several works have studied the organizational culture as another important factor (Aiman-Smith *et al.*, 2005; Adams *et al.*, 2006; Swart, 2006). According to these authors, innovations depend on the type of culture implemented in the organization. On that basis, firms that encourage risk-taking, creative behaviour or freedom to explore will facilitate innovation. But it is necessary to take sufficient control to manage innovation in an effective and efficient way.

Pizarro *et al.* (2007) found that the entrepreneurial culture does not have a direct influence on product innovation. Furthermore, Prajogo and Ahmed (2006) were unable to verify that incentives to innovate have a positive effect on product development. In this sense, Pizarro *et al.* suggest that the entrepreneurial culture is a contextual variable and does not have a direct effect on innovation, but rather has a moderating role on the relationship between the knowledge value of employees and innovation. Therefore, it is necessary to encourage an entrepreneurial culture in order to promote the exchange of knowledge among employees. Similarly, Prajogo and Ahmed indicated that human resources should be managed (by offering incentives to innovate) to encourage innovation.

It is also necessary to pay attention to organizational flexibility and responsibility for change, organizational centralization, organizational formalization, internal communication and so on, because they may encourage or hamper the innovation process (Boer and During, 2001; Benito-Torres and Varela-González, 2002; Adams *et al.*, 2006). For example, low formalization and centralization is appropriate at the beginning of the innovation process. In addition, it is necessary to take into account that the formalization should not be understood as an inflexible process, but rather as flexible planning.

Furthermore, based on Chesbrough and Teece (2003) and Adams *et al.* (2006), our study points out that different organizational forms will lead to different types of innovation.

Lastly, Damanpour (1991) showed that specialization (number of types of category within the firm) and functional differentiation (number of units below the level of senior management) have a positive and significant relationship with technical innovation, while centralization has a negative effect.

Damanpour (1991), and Muñoz and Cordón (2002) studied decentralization and specialization/complexity, asserting that these have a positive and significant influence on technical and organizational innovation, respectively – including new products and services, among others. It may be because, on the one hand, decentralization allows greater employee participation in decision-making, which facilitates organizational adaptations to change, since it is not necessary that all decisions are taken by senior management, thus speeding up the process; and, on the other hand, the complexity means having different organizational skills (derived from different units in the firm and the diversity of experts within them).

Damanpour (1991) and Muñoz and Cordón (2002) also analysed formalization, finding that it has no relevance to innovation. According to these authors, organizational procedures facilitate innovation in dynamic and uncertain environments, because in those cases it may be necessary to consider innovation within the formal definition of jobs in order to achieve flexibility.

Leiponen (2006) carried out research on the determinants of innovation, attempting to test empirically the positive influence of team skills and knowledge on the likelihood of implementing product innovation. She therefore considers the importance of routines and processes in the teams that are part of the firm. Similarly, Kyriakopoulos and De Ruyter (2004) indicated that these organizational routines have a positive effect on the creation of new products (referring to changes in products), but they add that such link has an inverted U-shape; hence, if the organization is too routine its effect on innovation will be negative.

In response to these results, it seems that too high a level of routine within firm may hamper creativity and the capability to perceive changes in the environment, though some level of structure allows for the defusing of potential situations of uncertainty, and improving the exchange and understanding of knowledge flowing around the organization. This explanation is in the same vein as the arguments on formalization presented by Muñoz and Cordón (2002).

With respect to competitors, and taking into account various theoretical perspectives (transaction costs theory, the resource-based view, evolutionary theory, and so on), Martínez *et al.* (2007) analysed organizational flexibility, both functional and strategic, finding a positive influence on product innovation. In this sense, these flexibilities are considered as part of organizational capital, because functional flexibility refers to the way that the firm distributes the different practices that exist within the organization (multipurpose teams, rotation tasks or telecommuting, for example), and strategic flexibility refers to the rapid response of the firm when faced with changes in the environment. The authors argued that it is necessary for employees to possess the ability to adapt to different tasks, and that capacity is achieved by the rotation of tasks and the development of versatile teams. Such practices allow a greater knowledge flow, which helps to improve the innovation process. In addition, firms involved in dynamic environments should adapt to the market and develop products that satisfy customers, underlining the important of strategic flexibility.

In contrast, Un and Cuervo-Cazurra (2004) showed that organizational-level integrative socialization, routine communication, and reward have a positive influence on product innovation. This is because socialization and routine communication facilitate the development of the capability to create knowledge by promoting understanding among individuals in different knowledge sets, and reward knowledge creation by influencing the willingness of individuals to interact and create knowledge.

With regard to communication, Akgün *et al.* (2007) also studied learning capability, taking into account knowledge transfer and integration. They considered the internal dispersal of knowledge via verbal and non-verbal communication, including conversation, dialogue, debate and interaction among individuals, both formal and informal, indicating empirically a positive relationship with product innovation. Nevertheless, when considering learning capability, Akgün *et al.* (2007) observed that a common identity and shared vision, interconnections in the activities of employees, and a climate of accepting new ideas and points of view, allowing individual knowledge to be constantly renewed, widened and improved, have a positive effect on product innovation. Also, Hegde and Shapira (2007) show that practices of knowledge management, promoting shared information and knowledge, and encouraging individuals stay in the firm, have a positive effect on product innovation.

With respect to process innovation, and in general terms, the work of Damanpour (1991) shows a positive association between this kind of innovation and organizational specialization and formalization.

In this same sense, Muñoz and Cordón (2002) analysed decentralization and complexity, finding a positive relationship with organizational innovation, in which they included new technologies, among others, that are considered as process innovation. On the other hand, they cannot show the negative influence of formalization on this kind of innovation. As stated previously, decision-making by different employees speed up the adaptation to changes, and the diversity of skills helps to achieve innovation.

Because Martínez *et al.* (2007) consider both product and process innovation when analysing innovation performance, it can be said that functional and strategic flexibility have a positive influence on process innovation.

On the other hand, structural capital includes technological issues. In our proposal, technological capital supposes a cornerstone in the model of innovation and intellectual capital presented (McEvily and Chakravarthy, 2002). Among the possible reasons for this, technological capital is intimately linked to the capability for technological innovation, since it includes both input measures related to innovation and prior technological knowledge historically possessed by the firm, which is labelled path dependence (Adner, 2002; Joia, 2004; Nerkar and Roberts, 2004).

Grant (1991) pointed out that innovations offer a temporary competitive advantage in emerging technological firms, where the speed of technological change is high. Therefore, in such cases, it is important to establish technological capabilities to carry out a direct current of innovations.

Specifically, Alegre-Vidaló *et al.* (2004) argued that R&D is one of the activities in the technological innovation process that has traditionally received more attention in the literature. In this sense, the efforts of a firm can be represented by its R&D expenditures, R&D intensity (ratio between R&D expenditure and turnover), participation in R&D projects with research institutions, or the formal existence of an R&D department. In this sense, many studies show a relationship between R&D expenditure and innovation (Achilladelis and Antonakis, 2001; Adams *et al.*, 2006; García and Mulero, 2007; among others). In addition, García and Mulero suggested that R&D expenditure may represent the innovative capability of the firm, because current R&D expenditure is often the result of prior R&D expenditure that led to successful performance.

In this sense, R&D expenditure is the traditional measure of formal innovation activities, and it is considered the most relevant input for knowledge creation and innovation (Huergo, 2006), assuming that the greater the investment in R&D, the higher the probability of success in technological innovation.

The other typical measure of technological effort is R&D personnel, assuming that larger firms probably have a larger proportion of personnel involve in R&D activities. This may be one of the reasons why, traditionally, the size of the firm was considered to be a determinant of innovations, alongside the logical economies of scale and scope in technological activities.

Nevertheless, as Huergo (2006) remarked, the commercialization of innovations requires complementary assets, and in particular the design of an interactive and dynamically efficient system for creating and transferring knowledge. In this sense, the interactions with other intellectual capital elements, such as information technologies, corporate values, trust, and relationships with the firm's external agents as customers, suppliers or allies, are essential for its success.

Hsieh and Tsai (2007) highlighted that technological capability – which includes technological knowledge, trade secrets and know-how engendered by R&D and other technology-specific intellectual property or patents – is the driving force of a firm's product innovation. Technological capital is a vital strategic resource for firms, especially high-tech firms, enabling them to stay in the lead position.

Based on these studies, we can formulate the second group of sub-hypotheses:

H2a: *The higher the structural capital endowments of the firm, the higher their product innovation.*

H2b: *The higher the structural capital endowments of the firm, the higher their process innovation.*

The role of relational capital in technological innovation

Finally, the present arguments investigate how different types of inter-organizational relationships contribute to the creation of new knowledge, and hence to improve a firm's technological innovation results.

Nowadays, many innovative firms spend very little on R&D, and yet they achieve successful innovations because of the knowledge and experience from a wide range of external sources (Laursen and Salter, 2006). Thus many studies suggest that knowledge from beyond the organizational boundaries of a firm is useful for innovation (Bossink, 2002; Chang, 2003; Phene *et al.*, 2006). More specifically, the latter work shows that external sources of knowledge are vital for a better product and process innovation. In addition, the authors state that the creation of radical innovations is a common function of external knowledge, and firms are able to access it.

We can expect that the interaction with different types of environmental agents, such as customers, suppliers, allies and so on, influence a firm's innovative performance. Knudsen (2007) summarized the motives by which firms engage in inter-firm collaborations: (i) the need for renewing the knowledge base through external learning; and (ii) the process of developing new or improved products is long and risky, hence the use of an external agent may save time and money for the firm.

The process of external learning extends over a long period of time. For this reason firms are motivated to establish permanent relationships with these external agents, as co-operative agreements. Furthermore, the accumulation of firm's relational experience within inter-organizational relationships is an important factor leading to the achievement of learning success.

As Huergo (2006) highlighted, the success in the commercialization of innovations involves complementary assets, not only intra-firms, but also inter-firm or relational ones. These complementarities would come from relationships with customers, suppliers, public and/or private research institutions, or even competitors in the product market.

Knudsen (2007) postulated that relationships with universities and research institutes, suppliers and competitors have a positive effect on innovative performance. But relationships with customers may have the opposite result. This may occur if the firm is too focused on the specific needs of a single group of customers, and therefore collaborations with customer in product development projects will have a negative effect on overall innovative performance.

The value held by the relationships that the firm maintains with the different agents in its competitive environment (mainly customers, allies, suppliers, plus other firms and institutions), or simply relational capital, constitutes a good source of information and knowledge-gathering for the firm. In this sense, the traditional view of innovation through sequential linear models of push-and-pull technology has evolved into a concept that sees innovation as a multi-actor process requiring high levels of interaction at both the inter- and intra-firm level (Adamides and Karacapilidis, 2006). This notion is closely related to relational capital. Following this vein, in recent years, one of the concepts that has been applied most frequently to determine the innovative capability of the firm is that of 'absorptive capability' (Cohen and Levinthal, 1990). These authors define this notion as the ability of the firm to recognize the value of novel external information, to assimilate it, and to apply it for commercial ends; hence absorptive capability becomes critical for innovation capability. In this sense, as Acedo *et al.* (2006) pointed out in their research about the dissemination and main trends of the RBV, a third main trend may be labelled 'the relational view'.

From the knowledge-based view, those processes and channels related to the close relations and collaboration agreements that can be found with suppliers or in strategic alliances can be understood as one of the main sources for organizational learning, and therefore also for technological innovation (Grant and Baden-Fuller, 2004). We must also take into account that firms can expand their ability to innovate to improve their relationships with customers, suppliers and other external groups (Tseng and Goo, 2005).

Furthermore, in the context of professional service firms, from a 'relational view', Natti *et al.* (2006) argued that effective relationships with customers lead companies to value creation in professional service firms. And empirical data show that relational capital has a positive influence on a firm's performance, in the context of biotechnology firms, measured as anticipated future sales.

With regard to relational capital, it is interesting to highlight the importance that the roles of networks, communities and research ties are assuming for innovation performance.

Similarly, Damanpour (1991) and Gallego and Casanueva (2007) asserted that external communication and co-operation, respectively, have a positive association with technological innovation. This is because involvement and participation by members in inter-organizational activities may lead to the development of innovative ideas.

Salman and Saives (2005), based on Powell *et al.* (1996), underlined clearly that organizational capability to innovate cannot be studied without considering external organizational relationships. In this sense, the authors went further and argued that it is essential that the firm is located in a central position for creating indirect links with other firms, and it thus has access to a wider range of activities and a greater chance of being located in positions of the greatest interest in order to obtain relevant information. Therefore, it is assumed that indirect networks usually foster the conditions for innovation and enable knowledge-sharing and transfer. Similarly, Almeida and Phene (2004) stated that networks with abundant amounts of knowledge offer greater opportunities for access to such knowledge, and thus greater opportunities to innovate.

When Damanpour and Gopalakrishnan (1998) studied sources of innovation, they highlighted mergers, acquisitions, joint ventures and strategic alliances, considering them to be alternative sources when introducing innovations. Because such sources are relationships that firms maintain with external agents, they are a part of relational capital.

In the same sense, Dyer and Singh (1998) indicated that alliance partners of the firm are, in many cases, the most important source of new ideas and information leading to innovation, because these inter-organizational relationships facilitate the exchange of knowledge. On the other hand, and based on other authors, McEvily *et al.* (2004) argued that research into alliances has explored effective mechanisms for learning and innovation. Specifically, King *et al.* (2003) suggested that alliances between small and large organizations lead to technological innovations because both organizations have complementary resources that may facilitate the success of technological innovation if they are combined.

Also, Swart (2006) pointed out that knowledge diversity makes it possible to create capability to innovate, and to achieve innovations it may be necessary to maintain long-term relationships involving major knowledge and information exchanges. Moreover, innovation can also be attributed to collaboration among organizations; this is relational capital.

With regard to agents, Adams *et al.* (2006) found widespread recognition in the literature that collaboration with suppliers and customers has a significant impact on the innovation process. Also, Chang (2003) observed relationships with associated, universities and R&D institutions, or participation in government projects, and their importance in technological innovation. Specifically, Blumentritt and Danis (2006) suggested that the demands of customers and competitive pressures influence innovation, so if a firm interacts with customers, understanding their needs and satisfying their aims, this will have a positive impact on technical innovation (Han *et al.*, 1998).

Various studies show the positive influence on product innovation of relationships maintained with customers and suppliers, when analysing collaboration or co-operation alliances (Chang, 2003; Díaz *et al.*, 2006; Gallego and Casanueva, 2007; Martínez *et al.*, 2007), because of the opportunity to access resources that enable firms to face more complex challenges

Similarly, Wu *et al.* (2007) found a positive relationship between relational capital, customers' and suppliers' interactions, and product innovation, since in this way a firm can gain complementary knowledge and skills and achieve a greater number of innovations, or can acquire key information and knowledge (Yli-Renko *et al.*, 2001) about technologies, markets and the needs of customers (Díaz and De Saá Pérez, 2007). However, Huergo (2006) did not agree that vertical co-operation in R&D, which also considers relationships with both customers and suppliers, is a relevant source of such innovation. It may be because Huergo based his work exclusively on R&D co-operation, an important dimension to process innovation rather than product innovation, since its main purpose is to improve procedures among different actors in the value chain.

With respect to relationships with competitors, Chang (2003) and Díaz *et al.* (2006) showed that they have a positive effect on product innovation, though relationships with customers and suppliers had only a minor effect, according to Díaz *et al.* (2006). However, Gallego and Casanueva (2007) indicated that product innovation is derived largely from the use of knowledge and information gathered from collaborations with other organizations that are not direct competitors; that is, they excluded potential competitors with opportunistic behaviour.

Various studies analyse relationships with universities, research institutes or organizations, and participation in European Union and/or government projects. In this sense, Chang (2003), Díaz *et al.* (2006) and Gallego and Casanueva (2007) showed a positive relationship between universities and/or technology centres and product innovation. In addition, Díaz *et al.* (2006) suggested that participation in European Union projects has a positive impact on the likelihood of carrying out product innovations, but that this relationship has a minor effect compared to relationships with customers or suppliers. This finding may be because vertical relationships are more common and closer to the business, and therefore they may offer greater value for the firm when carrying out innovations.

With regard to relations with partners, Stuart (2000) and Sampson (2007) indicated that such relations have a positive effect on product innovation. And, considering intellectual capital, Wu *et al.* (2007) found a positive relationship between relational capital, which includes partners' interactions, and product innovation. Thus external knowledge acquisition helps the achievement of innovations because complementary skills are shared. Moreover, Stuart (2000) suggested that innovative organizations possess key technological capabilities, so it is expected that the know-how gained from partners should support to the firm in developing new technology in the next period.

Hagedoorn and Duysters (2002) showed that strategic technology alliances (co-operation between firms in which at least one innovative activity is combined, or a technology exchange is carried out) are the main mechanism of firms that operate in technology-intensive sectors and want to acquire external innovative capabilities, which are considered new products.

Li and Atuahene-Gima (2002) analysed the specific relationship between different types of alliance and product innovation. They stated that relationship is positive if the alliance is for product development and negative if it is for marketing co-operation. The former has a positive relationship because of the existence of technical skills transfers between partners; and the latter has a negative relationship because of this type of alliance can divert limited resources, which are necessary for technological innovation, towards sales activities to support the partner instead of developing technical skills, thus hampering the achievement of product innovation.

Hayton (2005) argued that there is a positive relationship between corporate reputation and innovation, including new products and services. This is because a positive reputation reduces the risk perceived by potential agents when establishing a new interrelationship or alliance.

Finally, Tsai (2001) showed the positive relationship between a central network position and product innovation in relative terms. Our study finds that such a position is valuable at inter-organizational level, so it is assumed that the better the relationship with other firms or agents, the greater the product innovation because of possible exclusive information being available, and higher-level and better contacts have been made.

With regard to process innovation, Chang (2003), Díaz *et al.* (2006), Gallego and Casanueva (2007) and Martínez *et al.* (2007) found a positive effect between process innovation and relationships with customers and suppliers when they were considering collaboration or co-operation alliances. The achievement of process innovations may be a result of the importance of vertical links throughout the productive process.

More specifically, as has been explained in the previous subpropositions, Huergo (2006) showed how R&D co-operation with customers and suppliers is relevant to the likelihood of carrying out process innovation, since this kind of co-operation brings knowledge about procedures to be developed among different stakeholders.

As explained earlier regarding product innovation, Chang (2003), and Díaz *et al.* (2006) analysed the relationship maintained with competitors, pointing out that it has a positive effect on process innovation, though the relationship with customers and suppliers has only a minor effect, according to Díaz *et al.* (2006). Nevertheless, Gallego and Casanueva (2007) maintain that process innovation comes from collaboration with other organizations that are not direct competitors; that is, they excluded potential competitors displaying opportunistic behaviour.

As for universities, research institutes or organizations, participation in European Union and/or government projects, Chang (2003), and Díaz *et al.* (2006) returned to that issue. However, in this case, the relationship has a minor effect on the likelihood of carrying out process innovations that maintained relationships with customers or suppliers (Díaz *et al.*, 2006). And, more specifically, Barañano *et al.* (2005) suggested that the most significant relationships are with universities or research institutes. This may be because the agents, who are part of the production process of the firm, attempt to create or improve processes by using the relationship maintained with the firm.

Chang (2003), Díaz *et al.* (2006), and Gallego and Casanueva (2007) all indicated the existence of a positive relationship between universities and/or research institutes and process innovation, which also occurs at a lower level than in relationships with customers and suppliers (Díaz *et al.*, 2006), for the same reason as the earlier argument.

On the other hand, as was presented in the previous subproposition, Hagedoorn and Duysters (2002) referred to strategic technological alliances, indicating that these are the main mechanism for firms that operate in technology-intensive sectors and want to acquire external innovative capabilities, which are considered new processes.

With regard to product innovation, Hayton (2005) showed the existence of a positive relationship between corporate reputation and the number of new products, collected within the variable 'innovation'.

Thus, after the arguments submitted above, we can create the third group of subhypotheses:

H3a: *The higher the relational capital endowments of the firm, the higher their product innovation.*

H3b: *The higher the relational capital endowments of the firm, the higher their process innovation.*

5
Methodology

Now that the research has been described, in this chapter we shall address the different stages that are necessary to carry it out, providing details about the population and sample firms that were included in the study, the way that we approached them, and the measurement tools we employed for data gathering.

First, we explain the process for selecting the population of firms selected for empirical research, deciding to focus on high and medium-high technology industries located in Spain.

Next, the design process we followed to obtain appropriate measurement tools is described. These tools allowed us to access human, structural and relational capital, as well as the technological innovation results depicted in previous chapters, in order to undertake an exploratory and confirmatory analysis for intellectual capital blocks and product innovation performance. Bearing this purpose in mind, an 'ad hoc' multi-item survey questionnaire was developed, covering: (i) human capital assets; (ii) structural capital assets; (iii) relational capital assets; and (iv) product innovation performance.

Finally, in this chapter we also describe the different stages that have been followed in the information-gathering process, providing a research résumé sheet covering all the fieldwork carried out.

Population and sampling procedures

Following the arguments of Reed and DeFillippi (1990), as well as Godfrey and Hill (1995), the resource based view, as well as the knowledge based theory of the firm, depicts firms as a complex, deep and path-dependent reality that makes each organization a unique entity. This conception makes it necessary for researchers to face an important challenge: they must take a limited number of firms, conditioned by a similar environment and surrounding context conditions (that is, selecting firms from the same or similar industries and economic sectors) to determine how they differ from each other according to their internal resources and/or intellectual capital. This kind of clinical analysis should follow the methodological rules for fieldwork set down by Eisenhardt (1989) and Leonard-Barton (1992). Godfrey and Hill (1995) encourage researchers to develop new research methodologies and measurement tools in order to explore complex and unobservable organizational phenomena.

One of the first problems that must be dealt with when carrying out empirical research is to choose an appropriate population for the work. Then a representative sample must be taken from the chosen population. There have been several recent proposals regarding the mechanisms and research protocols that can ease this process (see, for example, Rouse and Daellenbach, 1999; King and Zeithaml, 2003).

The research protocol proposed by King and Zeithaml includes, as the first step for any empirical research, the setting of the research range via the selection of industries and firms. According to these authors, researchers must seek a homogeneous industrial context for the chosen population and sample.

In a similar way, Rouse and Daellenbach (1999) point out that, when assuming that unique resources and competences hold the potential for developing competitive advantages, there is no point in choosing large samples and cross-sectional studies to research such internal elements, because it will be very hard to discover all the effects related to time and history, industry and environmental characteristics, firm strategy, and the resource or capability that the empirical research is trying to analyse. Therefore, it is necessary to control possible sources of bias. To do so, Rouse and Daellenbach propose the selection of firms from the same industry, because in this way it ensures that firms in the population share markets of strategic factors and face the same industrial attributes that affect strategic decisions.

For Eisenhardt (1989), these tactics can improve the possibility of obtaining a prudent and reliable theory; that is, a theory that fits the data to a great extent, thus increasing the chance that researchers will make novel findings from the empirically obtained data.

According to Rouse and Daellenbach (1999), a sample selection process like that previously mentioned should provide important advances for research about firm strategy, because this field has reached a point at which detailed data are necessary, as well as focusing on organizational processes, strategies and implementation. These are the keys for gaining more integrated and useful knowledge about competitive advantage.

King and Zeithaml (2003) introduced some methodological steps to measure organizational knowledge, which would be useful to our research. These are:

- Definition of the arena: industry and firm choice.
- Protocol design: developing the researcher's knowledge.
- 'Ad hoc' questionnaire sent to managers in order to measure firm resources.

Choosing established industries with well-defined industry limits provides enough control to increase the probability of being able to identify an exhaustive repertory of knowledge resources that could be assessed by industry managers, providing perceptions about resource value and its contribution to a firm's strategic success.

To carry out deeper research, both empirical literature reviews and interviews with industry managers must be used in order to become acquainted with the main issues and specific terminology of the industry. To extend the interviews, it is possible to request feedback with regard to the clarity, general impressions, and the possibility of involving other managers and/or organizations in the survey.

Whenever possible, different managerial assessments from several management positions should be taken into account. For every knowledge resource identified in the industry, each manager must consider the degree to which his/her organization has an advantage or is at a disadvantage with regard to the firm's main competitors (on a Likert scale of 1–7). It is interesting to highlight the range of managerial perceptions offered about the role of knowledge resources on sustained competition, demonstrating the heterogeneous configurations of knowledge resources among firms.

This methodology provides new and important contributions to the resource based view by identifying and measuring manager perceptions about knowledge resources. Some key aspects are listed below:

- Research results show that the role of knowledge resources is different among industries and firms.
- Manager perceptions about organizational knowledge also changes significantly among firms.
- A wide range of knowledge resources provide competitive advantage among firms of the same industry.

This methodology complements and extends the efforts made by using the resource based view in the identification and measurement of firm resources. It provides an effective complement focused on intangible resources identified by industry experts. Its results also show that managers from different industries are able to provide adequate descriptions of organizational knowledge.

On the one hand, we can observe that the previous methodological rules proposed might well be adequate for our research, because they refer to intangible organizational factors. But on the other hand, we can assimilate technological innovation to competitive advantage, because, in the chosen industries, innovation represents probably the best way to obtain and sustain positions of competitive advantage.

Following the above-mentioned recommendations, our aim is to develop and validate a conceptual framework focused on 'an intellectual capital-based view of technological innovation'. Attaining our research aims will make it necessary to develop new measurement tools and the gathering of new data.

Alongside the already noted methodological advice, there is another important recommendation when choosing the sample and population for empirical research: looking for an appropriate manifestation or paradigmatic cases related to the studied phenomenon in order to select firms or examples to study. Authors deeply interested in qualitative data suggest a deeper treatment and analysis, and they recommend being particularly careful about this issue. Such authors remark on the relevance of aiming case selection towards paradigmatic cases (Eisenhardt, 1989; Gummesson, 1989; Yin, 1993). Nevertheless, something similar must be done when selecting industries to obtain the firm population and sample.

According to all the comments up to now, the purpose of selecting a population for the empirical research must be guided by two fundamental principles: (i) choosing firms that face similar competitive dynamics, which will be achieved by focusing on the firms of a specific industry or homogeneous set of industries; and (ii) choosing firms in which the phenomenon that is going to be studied would clearly be present, which can easily be recognized by firms when taking part in the research, allowing researchers to obtain relevant information for further analysis.

These selection principles led us to choose industries that are labelled as 'high and medium-high technology' as the population for our study, and to focus on firms from Spain. Though they have not previously been commented on, criteria such as feasibility and access to data are also critical when carrying out empirical fieldwork. Thus these criteria influenced the geographical selection of firms and industries for this research. And we added some additional conditions to be fulfilled by firms in order to take part in the desired population for the study. These conditions were orientated towards providing a clearer observation of the studied phenomenon, as well as to ease operational work to some extent.

Characteristics of high and medium-high technology manufacturing industries located in Spain

As we have noted previously, the choice of industry with which to carry out our empirical research was influenced by certain recommendations and comments based on the literature review, and in particular on the recommendations made by Rouse and Daellenbach (1999, 2002), highlighting the importance of focusing on a single kind industry or an industry setting in which firms shared similar environmental characteristics.

This is the case with Johnson *et al.* (2002), who, in general terms and based on Granstrand (1998), remarked that a key characteristic of knowledge-based firms – those whose market value derives from all forms of knowledge embedded to them, including human capital, technological knowledge accumulated by R&D activities, patents, hardware and software information technologies and so on – is their dependence on intellectual capital assets. It is necessary to select an industry with the relevant characteristics, based on knowledge assets, to carry out a suitable empirical analysis, where intellectual capital could play a key role.

More specifically, Leitner (2005) indicates that in high and medium-high technology industries, intangible assets play strategic roles. Focusing on research and development firms, Leitner's research analyses different intangible asset configurations, as well as their relations and internal effects, their main purpose being to increase innovation. Similarly, Huergo (2006) indicates that some of the more R&D-intensive industries are those related to R&D-intensive manufacturing.

Therefore, taking into account that our research relies on a 'knowledge-based view of the firm', and considering that, in R&D-intensive industries, knowledge is a key production factor, we are going to validate the model discussed in the previous chapter in high and medium-high technology manufacturing industries located in Spain (see Table 5.1).

The previous description of high technology and medium-high technology comes from an industry classification made according to R&D intensity in a set of countries including Australia, Belgium, Canada,

Table 5.1 High and medium-high technology manufacturing industries

CNAE-93 Classification

CNAE	Industries
	High technology manufacturing industries
244	Pharmaceutical industry
30	Computer materials and office machinery
321	Electronic components
32–321	TV, radio and communications sets
33	Medical instruments, precision instruments, optics and watchmakers' instruments
353	Aeronautical and spacecraft construction
	Medium-high technology manufacturing industries
24–244	Chemical industries – except pharmaceutical industry
29	Machinery and equipment
31	Machinery and electronic sets
34	Automotive industry
35–353	Naval, railway, motorcycle, bicycle and other transport construction

Source: INE (2009).

the United States, France, Italy, Japan, the Netherlands, the United Kingdom and Sweden, and later, Denmark, and weighted by its relative importance on national production among the set of countries. In this way, in a first stage, two provisional lists were obtained, segmenting industries into three sets of categories: high, medium-high and low technology-based industries (one list covers 1970–80, and the other 1980–95. The above distinctions were made taking into account the changes related to technology intensity among industries over this twenty-five-year period (INE, 2009)).

Later, in 2001, the OECD presented a new classification, as can be seen from Table 5.2. This time, classification refers to 'high technology services', according to R&D intensity calculated from two measures of production, namely production value and value added (INE, 2009).

Therefore, as Rouse and Daellenbach (1999, 2002) recommend, we are going to focus on these two sets of manufacturing industries, with similar characteristics, understanding that, because of their degree of complexity, they need a continuous effort in R&D, and a solid technological base (INE, 2009). Indicators for high technology were initially conceived as a results measurement and on the impact of R&D (INE, 2009).

In detail, according to the data of 2007, Spanish high and medium-high technology industries showed a turnover of 193,025 million euros,

Table 5.2 High and medium-high technology manufacturing industries net sales

CNAE	Industries	2006		2007	2006–7 changes (%)
		Industry net sales	%	Industry net sales	
	High technology manufacturing industries	**28,167,399**	**11.4**	**28,984,940**	**2.9**
244	Pharmaceutical industry	12,349,825	5.0	13,576,158	9.9
30	Computer materials and office machinery	761,170	0.3	737,314	–3.1
321	Electronic components	1,325,412	0.5	2,056,587	55.2
32–321	TV, radio and communications sets	5,130,085	2.1	4,710,229	–8.2
33	Medical instruments, precision instruments, optics and watchmakers' instruments	4,004,492	1.6	4,403,130	10.0
353	Aeronautical and spacecraft construction	4,596,415	1.9	3,501,522	–23.8
	Medium-high technology manufacturing industries	**152,188,816**	**62.1**	**164,040,846**	**7.8**
24–244	Chemical industries – except pharmaceutical industry	35,290,263	14.4	36,769,594	4.2
29	Machinery and equipment	29,920,144	12.2	32,127,676	7.4
31	Machinery and electronic sets	20,586,795	8.4	22,927,006	11.4
34	Automotive industry	58,729,142	24.0	63,369,610	7.9
35–353	Naval, railway, motorcycle, bicycle and other transport construction	7,662,472	3.1	8,846,960	15.5
	Both groups	**180,356,215**	**73.6**	**193,025,786**	**7.0**

Note: Sales amounts shown in thousands of euros.
Source: INE (2009).

with an approximate increase of 7 per cent in comparison to the previous year. Specifically, high technology manufacturing industries represented 15.02 per cent, whereas medium-high technology manufacturing industries represented 84.98 per cent of the total (see Table 5.2). Finally, the previous turnover of both industries represented 30.84 per cent of all manufacturing industries, since its total turnover rose to 625,888 million euros in 2007.

Taking into account employment rates, high and medium-high technology industries employed a total of 1,496,100 workers in 2007, representing 7.3 per cent of the total Spanish workforce. In detail, the total number of workers in high technology industries was to up 186,900, whereas the rate in medium-high technology industries was to up 751,000 workers.

In addition, during 2007, the total R&D expenditure for high and medium-high technology manufacturing industries was 4,683.8 million euros, representing 62.8 per cent of the total R&D expenditure in Spanish manufacturing industries (see Table 5.3).

The number of employees occupied by R&D activities in high and medium-high technology manufacturing industries was 53,816. This represents 30.1 per cent of the total number of employees occupied in R&D activities across all Spanish industries, and 29.7 per cent of all Spanish researchers.

Table 5.3 R&D indicators in high and medium-high technology manufacturing industries, 2007

Industries	Internal expenditures		R&D personnel			
	Total	%	Total		Researchers	
			Total	%	Total	%
High technology manufacturing industries	1,302,502	17.5	11,380.3	13.0	6,424.5	15.3
High and medium-high technology manufacturing industries	1,113,804	14.9	14,958.9	17.1	6,051.5	14.4
Both groups	**2,416,306**	**32.4**	**26,339.2**	**30.1**	**12,476.0**	**29.7**

Note: Expenditure shown in thousands of euros.
Source: INE (2009).

Table 5.4 Number of Spanish firms

	01/01/2007	01/01/2008	Change (%)
Total	3,336,657	3,422,239	2.6
Manufacturing	244,359	245,588	0.5
Building	488,408	501,056	2.6
Commerce	845,229	843,212	−0.2
Other services	1,758,661	1,832,383	4.2

Note: Percentage data on number of firms not supplied by INE.
Source: INE (2009).

To define the character of innovation –we consider an innovative firm to be one that has introduced in the last three years products to the market that are technologically new or improved, or new process technology, or improvements in its methods of production (INE, 2009).

Regarding the number of firms, according to the data provided by INE (2009), by 1 January 2008 there was a total of 3,422,239 firms in Spain, a rise of 2.6 per cent compared to the previous period. When looking at only firms involved in manufacturing activities, these represent a 7.2 per cent of the total of Spanish firms (see Table 5.4.). In detail, we have calculated that high and medium-high technology manufacturing industries accounted for 14.84 per cent of Spanish manufacturing firms in 2008 (see Table 5.6).

If we consider firm size, measured as the number of employees, Spanish firms can be characterized by their small dimensions. More than 1.7 million firms (51.3 per cent) have no employees, and 958,711 firms (28 per cent) have only one or two employees. Adding these two groups together, then 80 per cent of Spanish firms have fewer than three employees. And firms with twenty or more employees represent only 5.5 per cent of total of Spanish firms (see Table 5.5) (INE, 2009).

Taking a closer look at the employment data for high and medium-high technology manufacturing firms, statistics show that 94.5 per cent of firms have fewer than 50 employees. In general in Spain, only 2.8 per cent of high and medium-high technology manufacturing firms have 100 or more employees (see Table 5.6).

Additional sampling criteria

Bearing in mind the purpose of selecting a group of firms that shows homogeneous characteristics, we followed several criteria. As has already been mentioned, using the CNAE-93 codes, we selected the industries of high technology manufacturing (244, 30, 321, 32–321, 33, 353) and medium-high technology manufacturing (24–244, 29, 31, 34, 35–353).

Table 5.5 Firms by industry and number of employees

	Total	Manufacturing	Building	Commerce	Other services
Total	3,422,239	245,588	501,056	843,212	1,832,383
No employees	1,754,374	82,227	233,477	414,054	1,024,616
1–2 employees	958,711	62,465	129,565	263,851	502,830
3–5 employees	345,848	35,426	62,235	89,291	158,896
6–9 employees	160,460	21,283	30,911	38,616	69,650
10–19 employees	110,369	21,135	25,545	22,283	41,406
20 or more employees	92,477	23,052	19,323	15,117	34,985

Source: INE (2009).

Second, while Subramaniam and Youndt (2005) argue that it is more probable that firms with more than 100 employees have formalized innovation systems and R&D activities, because of the nature of Spanish industry, we decided to focus our empirical research on firms with 50 or more employees. This requirement ensures, on the one hand, that the sample will not suffer from a bias through the inclusion of great differences in firm size, and on the other, will allow us to show the different work groups of every firm, and to examine the intellectual capital assets that relate the three types of capital included in our theoretical framework (individual – human capital; organizational – structural capital; and inter-organizational – relational capital).

Third, for operational and technical reasons, we decided to carry out the empirical research in Spain.

To obtain the data from the existing firms, carrying out a preliminary census, and filtering organizations according to the previously mentioned criteria for sample selection, several databases were used, available online through the Complutense University in Madrid. From all the databases consulted, we eventually decided to use the information available in the SABI (System for Analyzing Iberian Balances) database. This includes data of Spanish and Portuguese firms from all industries, offering firm profiles, news and financial data, according to the firms' size, industry and location (among other approaches and browsing criteria).

Table 5.6 Number of firms included in high and medium-high technology manufacturing industries and number of employees

Firm size. Year 2008	CNAE-93								
	24	29	30	31	32	33	34	35	Total
Total	4,447	15,581	1,146	2,946	1,037	6,163	2,227	2,902	36,449
No employees	939	5,541	608	658	342	2,783	377	1.151	12,399
1–2 employees	853	3,374	328	556	171	1,769	321	581	7,953
3–5 employees	582	1,954	105	422	121	679	322	309	4,494
6–9 employees	454	1,340	55	302	105	321	295	218	3,090
10–19 employees	565	1,518	24	357	97	306	309	247	3,423
20–49 employees	550	1,271	14	391	118	206	290	246	3,086
50–99 employees	229	340	7	127	34	56	108	75	976
100–199 employees	116	137	2	69	22	25	75	32	478
200–499 employees	116	82	1	39	20	15	91	33	397
500–999 employees	32	14	2	17	5	2	21	3	96
1,000–4,999 employees	11	10	0	8	2	1	11	5	48
5,000 and more employees	0	0	0	0	0	0	7	2	9

Source: Based on INE (2009).

Starting with the information provided by the SABI database, a specific database was compiled for the fieldwork, covering those firms that fulfilled the previously mentioned criteria. The companies that did not provide complete contact information and those that were duplicated in the database were eliminated from our research database. The total number of valid firms to be contacted to take part in our survey in Spain was 1,270 firms.

Measurement of variables

In this section we show the measurement of variables in order to carry out the empirical testing of the hypotheses raised. Based on the literature review, we show the main constructs, dimensions and indicators proposed.

As was mentioned in the theoretical framework sections, in order to deal with the new challenges that appear in the identification and measurement of intellectual assets, in recent years some new approaches have emerged (Leitner, 2005), providing, as in the case of our independent variables, an 'intellectual capital-based view of the firm'.

In addition, as Kaplan and Norton (2004) remark, the correct assessment of intangible assets is very important, since managers might measure and manage the competitive position of their firm in a more suitable and simple way.

So, based on the theoretical and empirical literature review of intellectual capital and technological innovation phenomena, a set of indicators have been identified with the aim of carrying out the measurement of variables for our research. The above-mentioned indicators are included in several dimensions as well as in the different components or building blocks of intellectual capital and technological innovation.

At this point is necessary to emphasize that, because of the scarcity of research focused on this issue, many of the indicators proposed in the study have been obtained from researchers and analytical frameworks outside the so-called 'intellectual capital-based view', but devoted to examining specific facets related to human, structural or relational capital.

Special attention has been paid when attempting to develop these indicators, and when providing a coherent structure for the different intellectual capital blocks into dimensions, and then into indicators.

First, taking into account the dimensions of human capital (see Table 5.7) that were explained in the chapter devoted to intellectual capital, this construct can be characterized by the intrinsic characteristics of

102

Table 5.7 Dimensions and indicators of human capital

Dimension	Indicator	Authors
Education and training	Resources employed in employee training	Based on Snell and Dean (1992); Dodgson and Hinze (2000); Skaggs and Youndt (2004); Zárraga and Bonache (2005); Nieto and Quevedo (2005); Moon and Kym (2006); Wu *et al.* (2008); Egbetokun *et al.* (2009)
	Training inside the firm	Based on Snell and Dean (1992); James (2000); Zárraga and Bonache (2005)
	Employees with university degree	Based on Daellenbach *et al.* (1999); James (2000); Hermans and Kauranen (2005); Wu et al. (2008)
Experience and abilities	Employees' experience	Based on Carmeli and Tishler (2004); Youndt *et al.* (2004); Subramaniam and Youndt (2005); Reed *et al.* (2006)
	Valuable abilities	Based on Lepak and Snell (2002); Shaw *et al.* (2005); Nieto and Quevedo (2005); Youndt *et al.* (2004); Subramaniam and Youndt (2005); Reed *et al.* (2006); Pizarro *et al.* (2007)
	Development of new ideas and knowledge	Based on Lepak and Snell (2002); Youndt *et al.* (2004); Chen *et al.* (2004); Subramaniam and Youndt (2005); Reed *et al.* (2006); Pizarro *et al.* (2007); Wu *et al.* (2007)
Motivation	Employees satisfaction index	Based on Chen *et al.* (2004); Foo *et al.* (2006); Moon and Kym (2006)
	Employees' commitment and responsibility	Based on Carmeli and Tishler (2004); Zárraga and De Saá (2005)
	Internal promotion ratio	Huselid (1995)

Table 5.8 Dimensions and indicators of structural capital

Dimension	Indicator	Authors
Culture towards innovation	Cultural values promoted	Based on Russell and Russell (1992); Scott and Bruce (1994); Chen et al. (2004); Aiman-Smith et al. (2005); Prajogo and Ahmed (2006); Moon and Kym (2006); Alegre and Chiva (2007); Pizarro et al. (2007); Alegre and Chiva (2008)
	Shared system of values, beliefs and aims	Based on Carmeli and Tishler (2004); Chen et al. (2004); Youndt et al. (2004); Subramaniam and Youndt (2005); Moon and Kym (2006); Wu et al. (2007)
	Innovation and experimentation promotion	Based on Prajogo and Ahmed (2006); Pizarro et al. (2007); Akgün et al. (2007); Alegre and Chiva (2007); Egbetokun et al. (2009)
Management commitment	Employees' participation in decision processes	Based on Carmeli and Tishler (2004); Alegre and Chiva (2007); Akgün et al. (2007)
	Leadership and support for the innovation process	Based on Damanpour (1987); Boer and During (2001); Lloréns et al. (2005); Chen et al. (2004); Elenkov and Manev (2005); Akgün et al. (2007)
	Shared beliefs about the future	Prajogo and Ahmed (2006)
Management based on communication and information technology (CIT)	Processes, systems and organizational structures	Based on Youndt et al. (2004); Ordóñez (2004); Subramaniam and Youndt (2005); Reed et al. (2006); Cordón-Pozo et al. (2006); Wu et al. (2007); Akgün et al. (2007)
	Learning from the past	Based on Chen et al. (2004); Zárraga and De Saá (2005); Akgün et al. (2007)
	Communication, co-ordination and information diffusion	Based on Smith et al. (1994); Darroch and McNaughton (2002); Un and Cuervo-Cazurra (2004); Prajogo and Ahmed (2006); Moon and Kym (2006)
R&D internal efforts	R&D employees	Based on Darroch and McNaughton (2002); Yam et al. (2004); Chen et al. (2004); Moon and Kym (2006); Gallego and Casanueva (2007); Díaz-Díaz and De Saá-Pérez (2007); De Saá and Díaz (2007); Huergo (2006)
	R&D expenditures	Based on Cohen and Levinthal (1990); Dowling and McGee (1994); Li and Calantone (1998); Daellenbach et al. (1999); Tsai (2001); Chang (2003); Greve (2003); Chen et al. (2004); Nieto and Quevedo (2005); Díaz et al. (2006); Makri et al. (2006); Gallego and Casanueva (2007)
	R&D department	Based on Tippins and Sohi (2003)

firms' employees. Thus human capital includes nine items, organized into three dimensions – education and training; experiences and abilities; and motivation (see Appendix 1: Questionnaire at the end of the book).

Second, the dimensions and indicators of structural capital, which can play an important role in the technological innovation process (see Table 5.8), have been analysed through twelve items that were grouped into four dimensions (see Appendix 1: Questionnaire).

Third, regarding the dimensions and indicators within relational capital, which attempt to represent the structural guidelines of the relations maintained by the firm with different environmental agents, corporate reputation was also included in this block because of its role in empowering relational capital (see Table 5.9). Relational capital was examined by twelve items grouped into four dimensions (see Appendix 1: Questionnaire).

Fourth, the technological innovation measurement was based on a well-known typology: product and process innovation. Each of these innovations was measured by three items with their corresponding indicators (see Table 5.10 and Appendix 1: Questionnaire).

Finally, it is necessary to pay attention to the control variables included in our research. In this respect, after the intellectual capital and technological innovation empirical literature review (see Table 5.11) it seemed suitable to consider the two control variables of firm size and firm age.

Firm size was measured taking the total number of employees of the firm (Negassi, 2004; Reed *et al.*, 2006), and organizational age was captured as the number of years since the date the firm was legally registered.

Information sources and data gathering

This section explains the design process we followed to develop specific measurement tools for our empirical research, as well as the content of these tools.

Thus first we mention the information sources employed in the research, the kinds of data collected and used, and the method for carrying out the collection, which in this case has been a survey. After this, the stages and contents of the process for carrying out the questionnaire survey are explained, as well as the measurement scales that it finally contained, distinguishing between each set of intellectual capital assets present in the empirical framework designed for the research.

In this research we have accessed primary sources of information, obtaining the data directly, since there was no information available on a large number of firms regarding their different intellectual capital

Table 5.9 Dimensions and indicators of relational capital

Dimension	Indicator	Authors
Customer relationships	Market needs and trends	Based on Li and Calantone (1998); Yli-Renko *et al.* (2001); Tippins and Sohi (2003); Chen *et al.* (2004)
	Development of customer solutions	Based on Youndt *et al.* (2004); Subramaniam and Youndt (2005); Reed *et al.* (2006)
	Customer base	Based on Chen *et al.* (2004)
Supplier relationships	Development of solutions with customers	Based on Kotabe *et al.* (2003); Youndt *et al.* (2004); Subramaniam and Youndt (2005); Reed *et al.* (2006)
	Quality and design improvement with suppliers	Based on Kotabe et al. (2003); CIC (2003); Gallego and Casanueva (2007); Díaz-Díaz and De Saá-Pérez (2007)
	Supplier base	Based on Chen *et al.* (2004)
Ally relationships	Development of solutions with allies	Based on Youndt *et al.* (2004); Subramaniam and Youndt (2005); Reed *et al.* (2006)
	Quality and design improvement with allies	Based on Dodgson and Hinze (2000); Nieto and Quevedo (2005); Gallego and Casanueva (2007); Díaz-Díaz and De Saá-Pérez (2007)
	Ally base	Based on Chen et al. (2004)
Corporate reputation	Product and service reputation	Based on Dollinger *et al.* (1997)
	Management reputation	Based on Dollinger *et al.* (1997); Carmeli and Tishler (2004)
	Financial reputation	Based on Dollinger *et al.* (1997); Carmeli and Tishler (2004)

assets and innovation processes to fit a framework such as that presented in this book. Besides, using proxy variables might well have distorted our research.

Since we wanted to gather information about intangible assets that are internal organizational phenomena (apart from some specific relational assets), we took internal sources of information as being the only appropriate means for data gathering in our empirical research.

Table 5.10 Typology and indicators of technological innovation

Innovation typology	Indicator	Authors
Product innovation	Number of product innovations	Based on Miller (1987); Zahra and Covin (1993); Chandy and Tellis (1998); Tsai and Ghoshal (1998); Souitaris (2002); Li and Atuahene-Gima (2002); Wang and Ahmed (2004); Ordóñez (2004); Hayton (2005); Akgün *et al.* (2007); Jensen *et al.* (2007); Gallego and Casanueva (2007); Lin and Lu (2007); Wu *et al.* (2008)
	Percentage of total sales from new products	Based on Chandy and Tellis (1998); Souitaris (2002); Chen *et al.* (2004)
	New products versus firm products portfolio ratio	Based on the authors' original work.
Process innovation	Number of process innovations	Based on Zahra and Covin (1993); Souitaris (2002); Wang and Ahmed (2004); Hayton (2005); Sousa (2006); Gallego and Casanueva (2007); Lin and Lu (2007); Wu *et al.* (2008)
	New processes with reduction of production time and/or operations flexibility	Based on Alegre et al. (2005); Sousa (2006); Alegre and Chiva (2007)
	New processes with cost reduction	Based on Alegre *et al.* (2005); Sousa (2006); Alegre and Chiva (2007)

Table 5.11 Control variables

Item	Authors
Firm size	Tsai (2001); Tsang (2002); Tippins and Sohi (2003); Greve (2003); Skaggs and Youndt (2004); Youndt *et al.* (2004); Negassi (2004); Carmeli and Tishler (2004); Salman and Saives (2005); Reed *et al.* (2006); Díaz-Díaz and De Saá-Pérez (2007); Alegre and Chiva (2007); Pérez-Luño *et al.* (2007); García and Navas (2007); Jensen *et al.* (2007); Batjargal (2007)
Firm age	Li and Atuahene-Gima (2002); Youndt *et al.* (2004); Hermans and Kauranen (2005); Salman and Saives (2005); Reed *et al.* (2006); Díaz-Díaz and De Saá-Pérez (2007); García and Navas (2007); Batjargal (2007)

One of the main objectives of our work is to provide empirical evidence in order to make contributions to the 'intellectual capital-based view of the firm'. This has been the main reason for carrying out empirical fieldwork focused on quantitative data. Nevertheless, we must highlight that we are working in a field that is still quite new and not fully tested by empirical work, so this conditions the goals and results of this empirical research, focused on describing and confirming the use of measurement tools, and on verifying the concepts that have been depicted in our theoretical model.

There are several reasons for taking the survey method as an appropriate technique for data gathering. First, it is a suitable method for getting data of the type we needed, and makes it easier to access the intended information sources. Second, surveys allow us to attain a higher number of firms' responses compared with other research methods, such as in-depth interviews for example. Lastly, it can provide important economies of time and effort when carrying out the data gathering and its statistical treatment.

Designing the 'ad hoc' questionnaire

To prepare the questionnaire we used to collect quantitative data from the primary and internal sources chosen for our research, we followed a process divided into four phases: (i) literature review; (ii) design of the questionnaire in an initial version; (iii) pre-testing that preliminary version; and (iv) correcting and reframing the questionnaire to obtain a final version to be used for our fieldwork.

The first of these phases, questionnaire design, involved the reviewing of bibliographical references focused on the fields of RBV, intellectual capital and technological innovation, and entailed important challenges. First, this kind of literature is quite dispersed and heterogeneous in its nature. This can be explained, in a great extent, by the fact that this research stream is still at the 'in-development' stage (Dean and Kretschmer, 2007), showing multiple origins and theoretical roots, which complicates general agreement and definitions, despite enriching the discipline. Similarly, most of the contributions have only a theoretical character, or when addressing empirical research, they employ research methodologies based mainly on qualitative data. This makes it quite complex to use a survey as the base for data gathering taking as a starting point available questionnaires designed by other authors, since these are almost non-existent.

Regarding empirical research based on quantitative data analysis, we found once again the problem of heterogeneity in treatment. Each work is focused on different aspects of knowledge assets, and this usually forces researchers to use different measurement tools, and employ different questionnaire items.

The objective of the questionnaire to be used for fieldwork was to deal explicitly with the knowledge or intellectual capital assets and their roles in technological innovation that were present in the specific theoretical framework of earlier research (see the previous chapters of this book). Hence, because of the disparity between perspectives for analysis and the wide range of concepts that appear in this research field, and considering that none of the empirical proposals reviewed was fully adapted to the pursuit objective, we decide to design a specific questionnaire starting directly from:

- the theoretical arguments or examples that can be found in the explanations and reasoning of authors such as Subramaniam and Youndt (2005), Reed *et al.* (2006), or Alama Salazar (2008); and from
- the dimensions or partial concepts that, among all the previously mentioned works, were considered interesting and useful for the purpose of this empirical research.

For this reason, to different extent, all the previously mentioned empirical works can be considered as inspiring elements of the measurement tools that our questionnaire proposes, which will be detailed later.

The second phase we addressed in developing the questionnaire was writing an initial proposal, taking the measurement tools we considered to be most appropriate for the different intellectual capital assets and technological innovation outputs present in the theoretical model.

Third, before proceeding to the final version of the questionnaire, we carried out a pre-test with the initial version. This test consisted of submitting the proposed questionnaire to several expert academics in the field of intellectual capital and technological innovation, to be discussed, criticized and improved by their suggestions. Academics of the Manchester Institute for Innovation Research at Manchester Business School, University of Manchester (UK), Jaume I University (Spain), University of Salamanca (Spain) and Complutense University of Madrid (Spain) took part in this pre-test, as well as ten CEOs and industry experts.

Because the quantification of answers was not feasible in order to avoid a decrease in the response ratio, we decided to include scales showing 'in comparison with your main competitors' for those questions that might entail a higher subjectivity, as suggested by King and Zeithaml (2003).

Finally, once the suggestions of the academic and practitioner experts for modifying, adding or omitting questionnaire items were incorporated, as well as taking into account several corrections related to the writing and semantics of a number of elements, we produced a new version of the questionnaire to be sent to the population of firms that had been selected as candidates for the fieldwork of the research.

Similarly, after consultation with the experts, we also decided to change the planned procedure for contacting firms, abandoning our original idea of using e-mail or direct electronic contact, and adopted a contact protocol structured over several stages, with the invaluable help of consultants specializing in market and company research.

After the modifications were carried out, the questionnaire was structured as detailed in Table 5.12 and shown in full in Appendix 1.

In the following sections, we explain each of the groups of indicators that were developed as the starting point of the empirical research.

Specifically, given the strategic nature of intellectual capital, it seemed reasonable to gather data from managerial levels (Cabrita and Bontis, 2008), since managers are the employees with a wider vision of the firm's strategy.

Sample characteristics and statistical representativeness

As has been mentioned previously, because of the nature of the Spanish firms with regard to their size, the research population for our empirical analysis was shaped by high and medium-high technology manufacturing firms located in Spain and with 50 or more employees. The research population was obtained from the SABI database in 2007; a total of 1,270 firms.

Table 5.12   Structure of the questionnaire

Firm identification data	Firm name
	Identification of the person interviewed CNAE-93 code Number of employees Net sales Province Telephone number
Data about human capital assets	Nine items on 7-point Likert scale
Data about structural capital assets	Twelve items on 7-point Likert scale
Data about relational capital assets	Twelve items on 7-point Likert scale
Data about technological innovation	Six items on 7-point Likert scale

After sending the questionnaires, 251 firms indicated that they were willing to take part in our research (see Appendix 2: Sample firms), which gives a response rate of 17.07 per cent, with a sampling error of $+/-5.5$ per cent for a 95 per cent of confidence level.

Questionnaires were presented telephonically, using the services of a well-known Spanish firm devoted to market and company research studies: Merka Star. We provided this firm with our population database of 1,270 firms, and they contacted the companies individually. In a preliminary stage, Merka Star, with the assistance of members of the research group, carried out a pilot survey with twenty firms, to identify any mistakes in the initial version of the questionnaire.

Table 5.13 shows the empirical research résumé.

In terms of statistical inference, we can claim the sample to be considered as representative, because of the use of a method based on the random algorithm in Pascal used by Merka Star in order to decide whether a firm should be included in the sample or not.

In addition, in order to reaffirm the representation of the sample, we decided to carry out a comparative analysis between sample and population in terms of firm age and firm size, using several statistics showing trend and dispersion (see Table 5.14).

Considering firm age, most of the data in the sample are quite similar to those of the population, obtaining a p value = 0.212 when developing the test of average differences, with a confidence level of 95 per cent. Therefore it is possible to reject the null hypothesis about having a significant difference between the sample and the population.

Table 5.13   Research résumé

Target population	1,270 high and medium-high technology manufacturing firms
CNAE-93	24, 29, 30, 31, 32, 33, 34, 35
Firm size	Fifty or more employees
Geographic zone	Spain
Research unit	Firm
Data gathering	Telephone questionnaire
Sample size	251 valid questionnaires
Response rate	17.07 per cent
Sample mistake	$+/-$ 5.5 per cent
Sampling technique	Random algorithm in Pascal language
Statistical software	SPSS 17.0 and AMOS 7.0
Fieldwork	28 February to 5 May 2009
Respondents	Managerial level

Table 5.14 Sample representation

Statistics	Population firm age	Sample firm age	Population firm size	Sample firm size	Sample firm size*
Mean	25.51	26.78	166.10	233.43	201.89
Median	23	23	89	99	98
Standard deviation	15.329	16.87	289.811	628.97	340.65
Min	2	1	50	50	50
Max	84	108	3,362	11,571	4,914

Note: *Deleting top five firms by size (atypical cases).

Finally, considering firm size, it was necessary to delete the five largest firms (those with more than 5,000 employees), considering them to be atypical cases that distort the results, which can be detected in the extremely high figure for standard deviation (628.97). Having done this, after another average difference test we think that the sample is now representative, because it provides a p value = 0.052 for a confidence level of 95 per cent, rejecting once again the null hypothesis about a significant difference between sample and population.

6
Research Results

Taking into account the recommendations detailed in the previous chapter, in this one we show the main empirical results obtained from this research. Thus, through this chapter, we shall be able to test the causal relations we have posited in the book. Our explanations will be shown in three main steps: the first an introductory explanation, will be focused on a preliminary statistical analysis. The second is devoted mainly to obtaining a firm's intellectual capital measurement model from the acquired data, using exploratory and confirmatory factor analysis. And the third shows the results regarding the role of intellectual capital on product and process innovation outputs, tested in Spanish high and medium-high technology manufacturing firms.

In the preliminary analyses, we provide a first approach to the responses of these firms. To complete this objective we used descriptive statistics, showing which aspects, of those that were covered by the questionnaire, are most common in business practice, as well as in which of these firms that took part in the survey show more agreement or have important differences among them.

To design, develop and implement a measurement model of intellectual capital, exploratory and confirmatory factor analyses were the statistical techniques we chose for data processing. To develop our answers to such a complex phenomenon, we carried out four specific analyses. The first shows what kinds of assets can be included in human capital. The second is focused on structural capital, which includes intangible assets associated with technology management, as well as, with organizational structure, mainly informally. The third is devoted to the external relations maintained by the firm with its agents, as customers, suppliers or allies, among others. Finally, the fourth factor analyses report on the main product and process innovation outputs performed by the firms included in our sample.

In each one of the four exploratory factor analyses, in addition to pointing out assets or outputs that appear in the companies in our sample, we comment on which indicators, from all those included in the survey, allow us to discover if any of the identified assets is indeed performed by specific firms.

With the aim of validating the previous exploratory analysis, as well as obtaining statistically validated measurement models for intellectual capital blocks as well as product innovation outputs, we have run four complementary factor analyses for confirmatory purposes. These analyses will provide valid and reliable measurement models that can be found in high and medium-high technology manufacturing firms.

Preliminary analysis

This section introduces the reader to the research results. To do so, we provide a set of initial analyses regarding the frequencies of each group of intellectual capital blocks that were included in the employed questionnaire.

Descriptive statistics

As an initial approach to the analysis of the acquired data, we look at the main descriptive statistics (mean, median, mode, standard deviation, maximum value and minimum value) for each of the items included in the questionnaire. This analysis allows us to detect the most and least common activities that are performed in the firms to carry out each of the sets of processes that we have posited in the theoretical model for intellectual capital. This will also report on what activities show the widest differences when considering the firms of our sample.

The first group of descriptive analyses we provide corresponds to the nine questionnaire items that were related to operative human capital and its main dimensions. Table 6.1 shows the statistics for this analysis.

Taking into account education and training, the best firm endowments refer to personnel training (HC1 and HC2), both resources being considered as employed as specific training inside the firm. In second place, with a mean value of 4.54 over 7 (HC3), in general terms, the firms finally included in our research sample have a high level of personnel with university degrees. In addition, workers with education and training will be the best candidates for inclusion in internal promotion programmes.

Considering human capital endowments referring to personnel experience and abilities, firms included in our sample show high values of these compared to firms surveyed in previous research (mean values of 5.84/7 in the case of personnel experience (HC4), 5.45/7 for valuable abilities (HC5), and 5.28/7 for the development of new ideas and knowledge (HC6).

Regarding workers' motivation, as the third most common aspect of a firm's human capital, this shows high values in the case of the satisfaction index (5.46/7 HC7). Special attention must be paid to personnel commitment (HC8); a value of 5.94 over 7 shows the highest value/ endowment of human capital aspects of our sample firms. This item suggests that workers' commitment may be one of the key differential characteristics of human capital endowments in high and medium-high technology firms.

The second as third group of descriptive analyses regarding frequencies that we shall provide is the one that corresponds to the twelve questionnaire items related to operative structural capital and its main dimensions. Tables 6.2 & 3 show the statistics used to carry out this analysis.

Designing and developing an intellectual capital model for high and medium-high technology firms: exploratory and confirmatory analysis

In the following sections we shall show the results of the exploratory and confirmatory factor analyses. These allow us to answer the first question of our research: what is the nature and configuration of intellectual capital that can be found in business practice? Nevertheless, we must bear in mind that the answers we obtained – that is, the set of intellectual assets we found in our field study – correspond exclusively to the firms demonstrating the characteristics that define the population of our research.

Regarding the different types of intellectual capital assets that are included in this research, we shall follow a similar sequence for information processing, attempting to synthesize the group of issues that were included in the questionnaire into a reduced number of factors or observed assets. The steps we took were as follows:

(i) Test the convenience of carrying out the factor analysis through the KMO (Kaiser–Meyer–Olkin measure of sampling adequacy) index, Bartlett's test, the analysis of correlations matrix, and the communalities of the variables.

Table 6.1 Human capital assets – descriptive statistics

	HC1	HC2	HC3	HC4	HC5	HC6	HC7	HC8	HC9
N Valid	251	251	251	251	251	251	251	251	251
Missing	0	0	0	0	0	0	0	0	0
Mean	4.87	4.83	4.54	5.84	5.45	5.28	5.46	5.49	4.59
Median	5.00	5.00	5.00	6.00	5.00	5.00	5.00	6.00	5.00
Mode	5	5	5	7	5	5	5	6	5
Standard deviation	1.465	1.340	1.351	1.010	1.070	1.437	1.040	1.191	1.514
Variance	2.147	1.796	1.826	1.020	1.145	2.066	1.082	1.419	2.291
Min	1	1	1	3	1	1	3	1	1
Max	7	7	7	7	7	7	7	7	7

Table 6.2 Structural capital assets – descriptive statistics

	SC1	SC2	SC3	SC4	SC5	SC6	SC7	SC8	SC9	SC10	SC11	SC12
N Valid	251	251	251	251	251	251	251	251	251	251	251	251
Missing	0	0	0	0	0	0	0	0	0	0	0	0
Mean	5.32	5.39	5.57	4.94	5.74	5.60	5.81	5.05	5.04	4.13	4.22	4.78
Median	6.00	6.00	6.00	5.00	6.00	6.00	6.00	5.00	5.00	4.00	4.00	5.00
Mode	6	7	7	5	6	6	7	5	5	4	5	7
Standard deviation	1.454	1.391	1.317	1.460	1.181	1.210	1.345	1.541	1.555	1.688	1.696	2.249
Variance	2.115	1.934	1.735	2.133	1.395	1.465	1.809	2.374	2.419	2.851	2.878	5.058
Min	1	1	1	1	1	1	1	1	1	1	1	1
Max	7	7	8	7	7	7	7	7	7	7	7	7

Table 6.3 Relational capital assets – descriptive statistics

	RC1	RC2	RC3	RC4	RC5	RC6	RC1	RC2	RC3	RC4	RC5	RC6
N Valid	251	251	251	251	251	251	251	251	251	251	251	251
Missing	0	0	0	0	0	0	0	0	0	0	0	0
Mean	5.71	5.44	5.50	5.45	5.75	5.48	4.98	5.29	5.03	5.87	5.64	5.76
Median	6.00	6.00	6.00	6.00	6.00	6.00	5.00	6.00	5.00	6.00	6.00	6.00
Mode	7	7	7	6	7	6	5	6	7	6	6	7
Standard deviation	1.242	1.546	1.303	1.354	1.182	1.288	1.604	1.523	1.562	1.184	1.327	1.296
Variance	1.543	2.391	1.699	1.833	1.397	1.659	2.572	2.321	2.439	1.403	1.760	1.679
Min	1	1	1	1	1	1	1	1	1	1	1	1
Max	7	7	7	7	7	7	7	7	7	7	7	7

(ii) Determine the part of the variance contained in the original variables that can be explained by the obtained factor.

(iii) Analyse the original components matrix, and the rotated components matrix once a Varimax rotation has been applied, interpreting the obtained factors and showing their main component, and decide their groupings by providing an appropriate label for each as a separated concept.

(iv) Determine the reliability level of the obtained measurement scales for addressing each of the knowledge transfer processes found in the empirical evidence. To do this we used Cronbach's alpha coefficients.

(v) Carry out confirmatory factor analysis in order to build measurement models, testing the goodness of global, incremental and parsimony fit. In addition, determine the validity and reliability of the obtained models

In the paragraphs that follow we will show the results of the tests that have been carried out in order to determine whether it is appropriate to carry out a factor analysis.

Human capital

Starting with the group of questionnaire items relating to the human capital components within organizations, we carried out an analysis to identify the factors or latent phenomena present in those items. Factor analysis summarizes the information contained in a dataset of m variables for a reduced number of factors that will represent these variables, with a minimum loss of information.

As was explained in the previous chapter, human capital is covered by nine items. However, for the reasons already discussed, it is necessary to carry out exploratory factor analysis (EFA) with the aim of identifying the dimensions that compose human capital from a statistical point of view.

Before commenting on the attained results, it is necessary to examine the value of the determinant of the matrix of correlations, to Bartlett's test and to the KMO measure of sample adequacy. It is possible to say that the exploratory factor analysis is pertinent to the values they present (see Table 6.4): the matrix determinant value is close to zero, the KMO index is over 0.7, and Bartlett's test has a significance level of 0.000.

In the preliminary EFA developed, two factors or dimensions were obtained. Nevertheless, and based on a specialized empirical literature review and research experience, we were forced to obtain a third dimension,

Table 6.4 Human capital exploratory factor analysis

Human capital items	Component		
	1 E&T	2 Mot	3 E&A
HC2: Training inside the firm	0.888		
HC1: Employed resources in training activities	0.857		
HC3: Employees with university degree	0.670		
HC9: Use of internal promotion	0.599		
HC8: Commitment and responsibility		0.832	
HC7: Employee satisfaction index		0.804	
HC4: Employee experience			0.843
HC5: Employee valuable abilities			0.741
HC6: Development of new ideas and knowledge		0.479	0.481
Explained variance (%)	27.847	20.483	18.838
Accumulated variance (%)	27.847	48.330	67.168
Cronbach's alpha	0.788	0.720	0.680

Matrix correlations determinant		0.032
KMO index		0.802
Bartlett's test	Approximate chi-square	844.646
	df	36
	Significance	0.000

Note: E&T = Education and Training; Mot = Motivation; E&A = Experience and Abilities.

differentiating motivation from skills and experience, because of the importance attached to these.

So the extraction of the factors was forced to three and, in addition, a varimax orthogonal rotation was applied, to achieve a better adjustment (see Table 6.4).

Based on the results shown in Table 6.4, human capital is composed of three main factors or dimensions, with a percentage of accumulated variance of 67.168.

Factor 1: Education and Training (E&T)

This factor includes knowledge acquired thorough university education and specific training that employees have received inside the firm.

In addition, internal promotion is included, since this will only be achieved by the most prepared workers.

The percentage of explained variance is 27.847, being the dimension with a higher explained variance. Cronbach's alpha of 0.788 means that it fulfils the reliability analysis of the measurement scale.

Factor 2: Motivation (Mot)

The second dimension includes those aspects related to the employee's motivation regarding the firm. In this sense, it is possible to think that those employees who are satisfied and engaged with the firm will be those who are more motivated to reach targets.

The percentage of explained variance is 20.483, being the second dimension with a higher explained variance. Cronbach's alpha of 0.720 means that it fulfils the reliability analysis of the measurement scale.

Factor 3: Experience and abilities (E&A)

Finally, we gathered into the same factor the experience and skills of the firm's employees. This dimension includes knowledge that has been slowly acquired over time, resulting from the practice and the intrinsic ability of every employee. In Table 6.4 it is possible to observe that HC6 is similar in both Factors 2 and 3. Nevertheless, we have decided to consider it as part of Factor 3 because it is related to employee skills, though the development of new ideas and knowledge might be a consequence. Furthermore, if it were decided to delete it, the KMO index of human capital would reduce to 0.779 and Cronbach's alpha to 0.655, in return for a slight improvement in the total percentage of accumulated variance (70.032).

The percentage of explained variance is 18.838, being the dimension with a lower explained variance. Cronbach's alpha of 0.680 is very close to the desired 0.7.

Once the exploratory factor analysis (EFA) had been carried out, the next step was the confirmatory factor analysis (CFA) to observe the unidimensionality of the measures of every dimension of human capital (for first order CFA, see Figure 6.1 and Table 6.5), and to check that the obtained dimensions fit with human capital and not with another construct (second order CFA).

Regarding the model fit (whether it be global, incremental or regarding parsimony), we can say that, according to the obtained index values, it is good. The mean quadratic approach error, with a value of 0.077, is located between the acceptable levels of 0.05 and 0.08 (Browne and Cudeck, 1989). Additionally, the global fit index (GFI) presents a

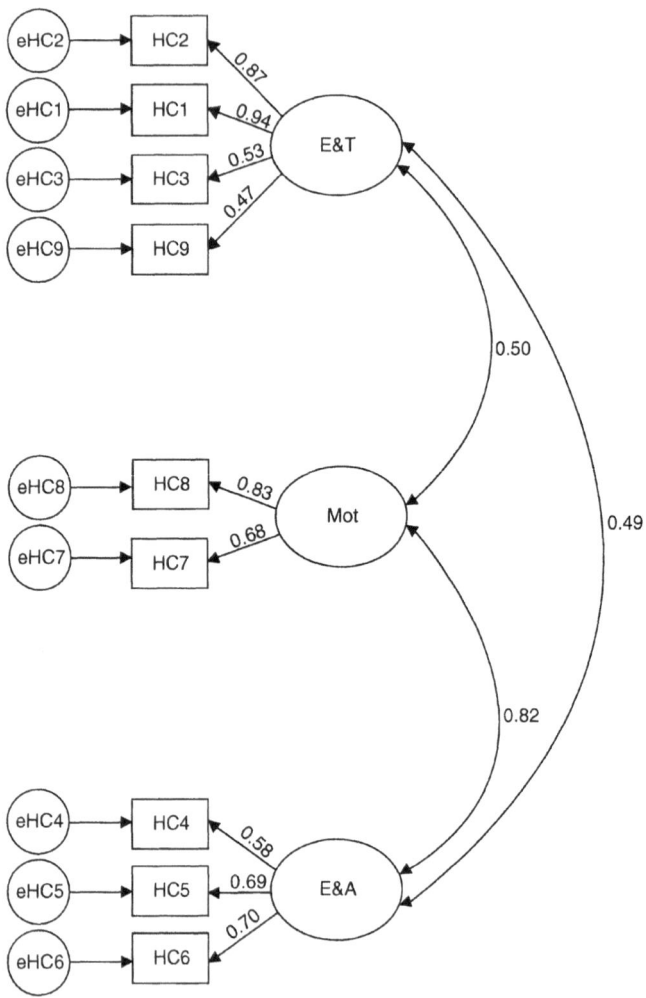

Figure 6.1   First order confirmatory factor analysis model of human capital

value over the minimum value for acceptance of 0.9. The values AGFI, CFI and IFI are close to one, which indicates once again that our data is a good fit. Finally, the value of the normal chi-square (or taking its relation to the grades of freedom shown by the model) is 2.464,

Table 6.5 Human capital first order model fit

Overall fit index
RMSEA
0.077

Comparative fit indexes			
GFI	AGFI	IFI	CFI
0.952	0.910	0.958	0.957

Parsimony fit index
Normal chi-square
2.464

Note: RMSEA = Root mean square error of approximation; GFI = Global fit index; AGFI = Adjusted goodness of fit index; IFI = Incremental fit index; CFI = Comparative fit index.

Table 6.6 Human capital first order regression weights (not standardized)

			Estimate	SE	CR	*p* value
HC9	←	E&T	1.000			
HC3	←	E&T	1.000	0.162	6.184	***
HC1	←	E&T	1.918	0.247	7.772	***
HC2	←	E&T	1.617	0.209	7.746	***
HC7	←	Mot	1.000			
HC8	←	Mot	1.401	0.155	9.028	***
HC6	←	E&A	1.000			
HC5	←	E&A	0.733	0.085	8.621	***
HC4	←	E&A	0.576	0.77	7.518	***

Note: SE = Standard error; CR = Critical ratio. $p < 0.01$.

remaining within the advisable levels that are required of this index (ranging between 1 and 3).

With regard to the unstandardized regression weights (see Table 6.6), it is clear that critical ratios (CR) are greater than 2 (a value in which the standard error is greater than zero) and are significant; that is, that the regression weight of every variable with regard to its factor is significantly different at a level of 0.001.

As can be deduced from the results, in the model all the regression coefficients of the measures are statistically different from zero and the *p* value of the chi-square is greater than 0.05, so the null hypothesis,

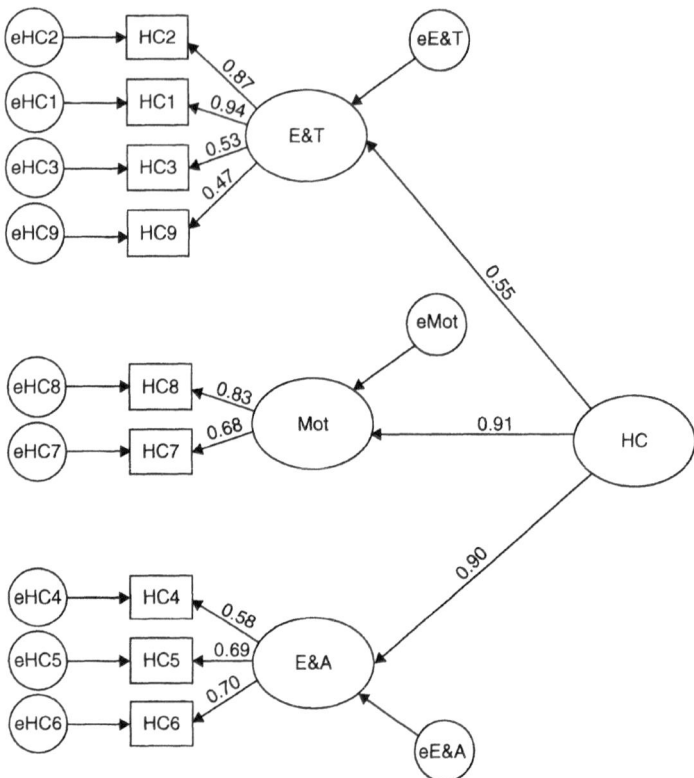

Figure 6.2   Human capital second order confirmatory factor analysis

which states that the estimated correlations matrix differs from the observed one only as a result of sampling error, cannot be rejected.

As for the second order CFA, an analysis was carried out (see Figure 6.2) with the aim of checking that the dimensions proposed for human capital fit this and not another construct.

In the same way as it was done for the first order CFA, the indicators were analysed to assess the model fit, which possess similar levels to the first order human capital model.

As for the regression weights, Table 6.7 shows that they are significant and have a few critical ratios greater than 2, as we found in the model of the first order. So the model's fit is appropriate and it can be asserted that human capital dimensions are part of this kind of capital.

Table 6.7 Human capital second order regression weights (not standardized)

			Estimate	SE	CR	*p* value
E&T	←	HC	1.000			
E&A	←	HC	2.302	0.462	4.979	***
Mot	←	HC	1.637	0.344	4.763	***

Alongside reliability analysis, it is necessary to determine the validity of the measurement model. We are referring to a situation in which the indicators measure precisely what they are supposed to be measuring. To carry out this task, four types of scale validity are used:

(i) *Concept validity*. The objective here is to assess whether a latent concept is related to the rest of the observed variables in a consistent way, according to the predictions made by the theoretical model (Lehmann *et al.*, 1999). To summarize, it is about determining, in our case, if we are indeed measuring a kind of capital assets. A way to assess that kind of validity is through alternative approaches, such as convergent validity and discriminant validity.

(ii) *Convergent validity*. This refers to the case in which a certain measure converges with the rest of the measures in a same model and all these measurements are part of the same construct. One way to determine this kind of validity empirically is by assessing the substance of the standardized factorial regression coefficients between the whole set of items and the corresponding latent variable. In this way, those coefficients that are statistically significant with a level of confidence of 95 per cent (t values greater than 1.96) and higher than 0.5, will show convergent validity. As can be seen from previous tables, the model we have used fulfils this condition easily.

(iii) *Discriminant validity*. According to this concept, a construct or latent variable should be sufficiently different from the rest of the constructs in order to justify its existence as an isolated individual item (Lehmann *et al.*, 1999). Because it is not possible to corroborate this kind of validity with objective measures related (or unrelated) to the proposed model, we have decided to carry out an alternative test for discriminant validity, similar to that described by Anderson and Gerbing (1988). This test takes the confidence interval of the correlation between each pair of critical dimensions (in our case we have one only pair) and checks that it does not contain the value 1. If this value is not present in the mentioned interval, it would be demonstrable that the correlation between the two dimensions is not close to 1, so it can be assumed that the pair of dimensions we have taken represent different latent concepts. The interval (0.4; 0.665) indeed demonstrates this.

(iv) *Criterion validity*. This refers to the effectiveness level we can expect when attempting to predict a variable starting from an existing measurement. In this case we are asking whether the proposed measures for a given concept exhibit generally the same direction and magnitude of correlation with other variables already measuring that concept that are accepted within the social science community. That is, this kind of validity is focused on the ability of a scale to reflect whether relationships with other variables – criteria variables – are indeed in accordance with to the theoretical predictions. Because of the lack of data available to carry out any test for this type of validity, we have decided not to address it in our research.

Reliability exists when measurements are seen to be consistent. In our research, reliability or the absence of random measurement errors for the latent variables, has been tested by analysing the internal consistency of the items used for their definition. In this case, they have been calculated by using a compound coefficient of reliability, an alternative to the Cronbach's alpha coefficient, because it is more appropriate in this case as it takes into account the different weights of each variable, without depending on the number of attributes considered for each dimension (Vandenbosch, 1996). These values are shown in Figure 6.2. As will be seen, both dimensions exceed the minimum required level of 0.7.

Structural capital

Dealing with the group of questionnaire items that aimed to analyse the structural capital components within the firm, we carried out a second exploratory factor analysis, in order to identify the factors or latent phenomena present in these items.

As we explained previously, structural capital was covered by twelve items, gathered into three dimensions. We needed to carry out the EFA in an attempt to identify the dimensions that compose structural capital (which includes organizational and technological assets) from a statistical point of view.

In a preliminary stage, it is necessary to concentrate on the value of the determinant of the correlations matrix, Bartlett's test and the

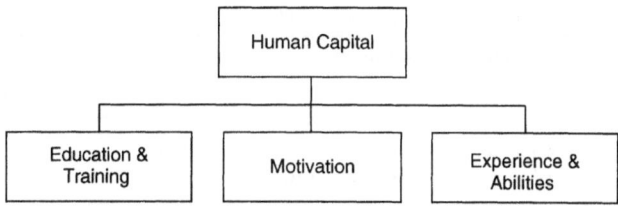

Figure 6.3 Observed model for human capital

Kaiser–Meyer–Olkin (KMO) measure of sample adequacy. It is possible to state that the exploratory factor analysis is pertinent for the values they present (see Table 6.8): the matrix determinant value is close to zero, the KMO index is over 0.7, and the Bartlett's test has a significance level of 0.000.

As can be seen from Table 6.8, three factors or dimensions were obtained using EFA. In Chapter 2 we defined four main dimensions for structural capital: (i) culture towards innovation; (ii) management commitment; (iii) CIT-based management; and (iv) R&D internal efforts. However, EFA results show how dimensions 1 and 2 can be aggregated into a unique factor or dimension, representing cultural values as well as managements' commitment towards innovation.

Based on the results shown in Table 6.8, structural capital is composed of three main factors or dimensions, with a percentage of explained variance of 72.451.

Factor 1: Cultural values and management commitment towards innovation (CUCO)

This factor refers to the values and cultural beliefs established within the firm as well as to managements' commitment towards the innovation process. As was observed in Chapter 4, it seems that management support and leadership have a key part to play in the effective implementation of cultural values, being included as a unique factor or dimension.

The percentage of explained variance is 32.786 – the dimension with a major power of explanation. Similarly, a Cronbach's alpha of 0.908 means it fulfils the reliability analysis of the measurement scale.

Factor 2: R&D internal efforts (R&D)

This factor represents the knowledge created and developed by means of internal creative developments and procedures, and through activities developed by R&D employees, as well as by all the internal

Table 6.8 Structural capital exploratory factor analysis

Structural capital items	Component		
	1 CUCO	2 R&D	3 CIT
SC2: Shared values and beliefs towards innovation	0.838		
SC5: Managers' support and leader innovation process	0.808		
SC3: Promotion of experimentation and innovation	0.805		
SC1: Creativity, innovation and development of new ideas are shared cultural values	0.737		
SC4: Employees involved in importantdecision-making processes	0.737		
SC6: Managers have shared beliefs about the future of the firm	0.737		
SC10: Percentage of R&D employees above the mean of the industry		0.887	
SC11: R&D expenditure/net income about the mean of the industry		0.884	
SC12: Existence of a formalized R&D department		0.769	
SC7: Use of CIT in co-ordination, communication and information activities			0.892
SC8: Learning by past experiences using CIT			0.845
SC9: Major part of organizational knowledge is stored in databases, intranet, etc.			0.620
Explained variance (%)	32.786	20.748	18.917
Accumulated variance (%)	32.786	53.534	72.451
Cronbach's alpha	0.908	0.838	0.804

	Matrix correlations determinant	0.000
	KMO index	0.865
	Approximate chi-square	1946.038
Bartlett's test	df	66
	Significance	0.000

Note: CUCO = Cultural values and management commitment towards innovation; R&D = Research and development internal efforts; CIT = Use of communication and information technology in management activities.

resources dedicated to R&D activities, and the existence of a formalized R&D department.

The percentage of explained variance is 20.478, being a second dimension with a higher explained variance. Similarly, a Cronbach's alpha of 0.838 means that it fulfils the reliability analysis of the measurement scale.

Factor 3: Use of CIT in management activities (CIT)

Finally, within this factor are included those aspects related to technological applications that are used to treat data about management and administration activities.

The percentage of explained variance is 18.917, thus being the dimension with a lower explained variance. And a Cronbach's alpha of 0.804 means that it fulfils the reliability analysis of the measurement scale.

Following the same steps as before, once the exploratory factor analysis (EFA) was completed, the next step was to carry out the confirmatory factor analysis (CFA) to observe the unidimensionality of measures of every dimension of human capital (for first order CFA, see Figure 6.4 and Table 6.9) and to confirm that the obtained dimensions fit structural capital and not to another construct (second order CFA).

Based on the values shown in the model, we can affirm that, according to the obtained index values, the model fit is good. The mean quadratic approach error, with a value of 0.084, is close to the acceptable levels of 0.05 and 0.08 (Browne and Cudeck, 1989), and the global fit index (GFI) presents a value greater than 0.9, the minimum value for acceptance. The values AGFI, CFI and IFI are high and close to one, so point out once again that we have a good fit. Finally, the value of the normal chi-square (or taking its relationship with the degrees of freedom shown by the model) presents a value of 2.746, remaining within the advisable levels required of this index (in the range between 1 and 3).

Additionally, with reference to the unstandardized regression weight (see Table 6.10), it reflects that critical ratios (CR) are greater than 2 (a value in which the standard error is greater than zero) and are significant; that is, that the regression weight of every variable with regard to its factor is significantly different by 0.001.

Based on the results shown in the model all the regression coefficients of the measures are statistically different from zero and the *p* value of the chi-square is greater than 0.05, so the null hypothesis, which states that the estimated correlations matrix differs from the observed one only as a result of sampling error, cannot be rejected.

As for the second order CFA, an analysis was carried out (see Figure 6.5) with the aim of checking that the dimensions proposed for structural capital fit this construct and not another. In the same way as to calculate the first order was calculated, the indicators were analysed to assess the model fit, which possesses similar levels to the first order human capital model.

Regarding the regression weights, Table 6.11 shows that they are significant and that they have a few critical ratios greater than 2, which we also

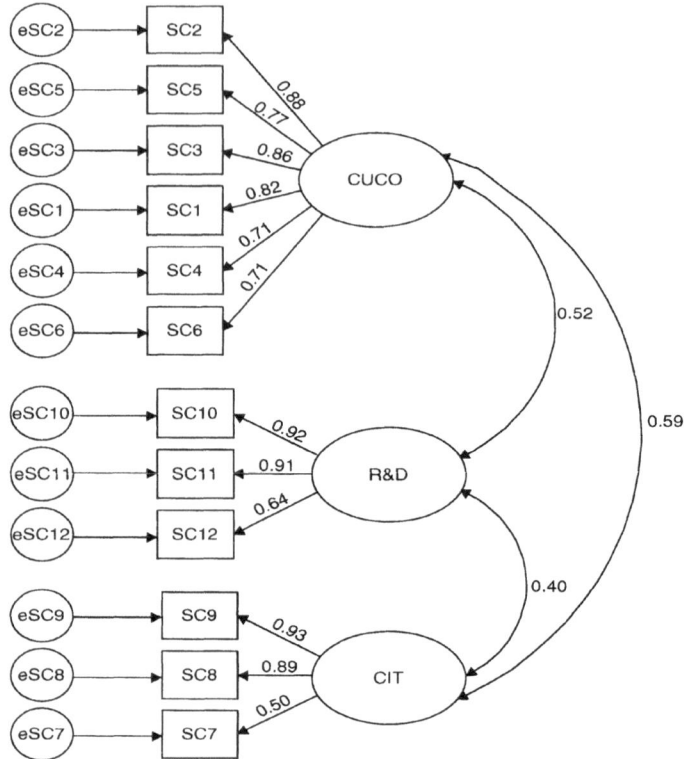

Figure 6.4 First order confirmatory factor analysis model of structural capital

found in the model of the first order. So the model's fit is appropriate and it can be asserted that structural capital dimensions are part of this kind of capital.

Taking into account the reliability analysis, it is necessary to determine the validity of the measurement model. As we noted earlier, four types of scale validity are used: concept validity; convergent validity; discriminant validity; and criterion validity.

- *Concept* validity can be tested through *convergent* validity. It makes reference to the case in which a certain measure converges with the rest of the measures in the same model and all of these measures are part of the same construct. One way to determine this kind of

Table 6.9 Structural capital first order model fit

Overall fit index
RMSEA
0.084

Comparative fit indexes			
GFI	AGFI	IFI	CFI
0.915	0.870	0.954	0.954

Parsimony fit index
Normal chi-square
2.746

Table 6.10 Structural capital first order regression weights (not standardized)

			Estimate	SE	CR	*p* value
SC6	←	CUCO	1.000			
SC4	←	CUCO	1.210	0.114	10.656	***
SC1	←	CUCO	1.391	0.114	12.250	***
SC3	←	CUCO	1.327	0.103	12.872	***
SC5	←	CUCO	1.061	0.092	11.536	***
SC2	←	CUCO	1.438	0.109	13.185	***
SC12	←	R&D	1.000			
SC11	←	R&D	1.078	0.094	11.428	***
SC10	←	R&D	1.084	0.095	11.438	***
SC7	←	CIT	1.000			
SC8	←	CIT	2.017	0.243	8.313	***
SC9	←	CIT	2.135	0.257	8.307	***

validity empirically is by assessing the substance of the standardized factorial regression coefficients between the whole set of items and the corresponding latent variable. Those coefficients that are statistically significant to a confidence level of 95 per cent (t values greater than 1.96) and higher than 0.5, will show convergent validity. As we can see from previous tables, the model we have used in our case fulfils this condition easily.

- *Criterion* validity makes reference to the effectiveness level we can expect when trying to predict a variable starting from an existing measurement. In this case we asked whether proposed measures for a given concept exhibited generally the same direction and magnitude of correlation with other variables as measures of that concept

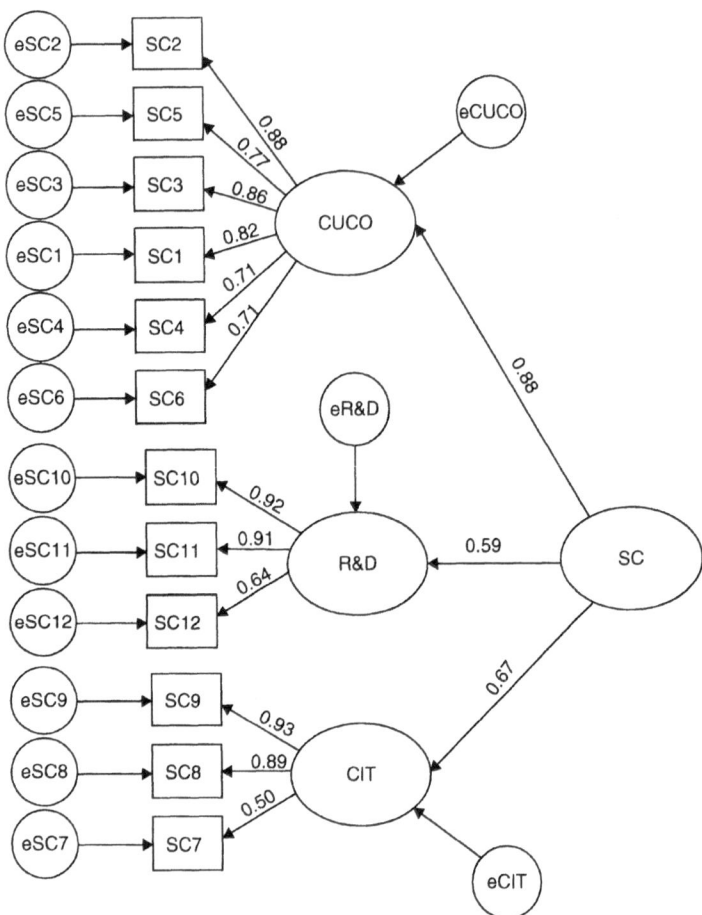

Figure 6.5 Structural capital second order confirmatory factor analysis

already accepted within the social science community. That is, that this kind of validity is focused on the ability of a scale to reflect whether relationships with other variables – criteria variables – are indeed in accordance with theoretical predictions. Because of the lack of data available to carry out any test for this type of validity, we have decided not to address this in our research.

Table 6.11 Structural capital second order regression weights (not standardized)

			Estimate	SE	CR	p value
CUCO	←	SC	1.657	0.330	5.021	***
R&D	←	SC	1.870	0.368	5.085	***
CIT	←	SC	1.000			

SE = Standard Error of Regression Weight
CR = critical ratio for Regression Weight
P = Level of Significance for Regression Weight

Jointly with the *validity* test, we needed to test reliability. In our research, reliability or the absence of random measurement errors for the latent variables has been tested by analysing the internal consistency of the items used to define these. They have been calculated through a compound coefficient of reliability, an alternative to the Cronbach's alpha coefficient. We considered it to be more appropriate in this case because it takes into account the different weights of each variable, without depending on the number of attributes considered for each dimension (Vandenbosch, 1996). Thus both dimensions are greater than the minimum required level of 0.7.

Relational capital

Finally, we analysed the last group of questionnaire items that attempted to show the relational capital components taking place in the relationships maintained by the firm with its key environmental agents, such as customers, suppliers, allies and so on. We carried out a factor analysis, to identify the factors or latent phenomena present in the questionnaire items.

As we explained earlier, relational capital was covered by twelve items, gathered into four dimensions. In order to test its empirical consistency, it was necessary to carry out the EFA with the aim of identifying the dimensions that compose the relational capital from a statistical point of view.

Before commenting on the final results, it is necessary to pay attention to the value of the determinant of the correlations matrix, Bartlett's test and the Kaiser–Meyer–Olkin (KMO) measure of sample adequacy. It is possible to say that the exploratory factor analysis is pertinent for the values they present (see Table 6.12): the matrix determinant value is close to zero, KMO index is over 0.9, and Bartlett's test has a significance level of 0.000.

In the EFA developed, a total of three factors or dimensions were obtained, and a varimax orthogonal rotation was applied, to achieve a better adjustment (see Table 6.12).

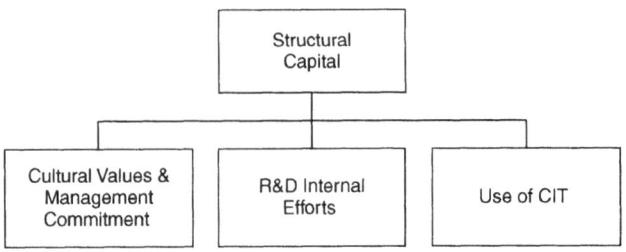

Figure 6.6   Observed model for structural capital

Based on the results shown in Table 6.12, relational capital is composed of three main factors or dimensions, with a percentage of explained variance of 73.195.

Factor 1: Customer and supplier relationships (C&S)

This factor is based on knowledge derived from the relations the company has with its customers and suppliers, to develop solutions and make improvements in its procedures. So those agents that maintain vertical relationships with the firm are grouped together.

The percentage of explained variance is 26.488 – so being the dimension with a higher explained variance. Similarly, a Cronbach's alpha of 0.866 means that it fulfils the reliability analysis of the measurement scale.

Factor 2: Relationships with allies (ASS)

This factor is focused on the relations the organization maintains with its allies, having a similar aim and procedures to the previous factor.

The percentage of explained variance is 23.669, being the second dimension with a higher explained variance. Similarly, a Cronbach's alpha of 0.901 means it fulfils the reliability analysis of the measurement scale.

Factor 3: Corporate reputation (REP)

In this factor, corporate reputation is represented by different aspects related to external agents, which will influence day-to-day business activities. Basically, it considers: (i) reputation regarding product and service quality; (ii) management reputation; and (iii) financial reputation.

The percentage of explained variance is 23.038, being the dimension with a lower explained variance. Similarly, a Cronbach's alpha of 0.900 means it fulfils the reliability analysis of the measurement scale.

Table 6.12 Relational capital exploratory factor analysis

Relational capital items	Component		
	1 (C&S)	2 (ASS)	3 (REP)
RC4: Development of solutions with suppliers	0.776		
RC2: Development of solutions with customers	0.714		
RC5: Quality and design improvements with suppliers	0.673	0.473	
RC1: Market necessities and trends	0.662		
RC3: Customer base	0.620		0.420
RC6: Suppliers' base	0.605	0.456	
RC7: Development of solutions with allies		0.833	
RC8: Quality and design improvement with allies		0.829	
RC9: Allies base		0.789	
RC12: Financial reputation			0.877
RC11: Management reputation			0.819
RC10: Product and service quality reputation			0.806
Explained variance (%)	26.488	23.669	23.038
Accumulated variance (%)	26.488	50.156	73.195
Cronbach's alpha	0.866	0.901	0.900

	Matrix correlations determinant	0.000
	KMO index	0.912
Bartlett's Test	Approximate chi-square	1995.189
	Df	66
	Significance	0.000

Note: C&S = Customer and supplier relationships; ASS = Relationships with allies; REP = Corporate reputation.

Once the exploratory factor analysis (EFA) was complete, the next step was to carry out confirmatory factor analysis (CFA) to observe the unidimensionality of the measures of every dimension of relational capital (first order CFA, see Figure 6.7 and Table 6.13) and to check that the obtained dimensions fit to relational capital and not to another construct (second order CFA).

Regarding the model fit, we can confirm that, according to the obtained index values, this is good. Tthe mean quadratic approach error, with a value of 0.081, is located between the acceptable levels of 0.05 and 0.08 (Browne and Cudeck, 1989). And the global fit index (GFI) has a value greater than the minimum value for acceptance of 0.9, and the values AGFI, CFI and IFI are close to one. Finally, the

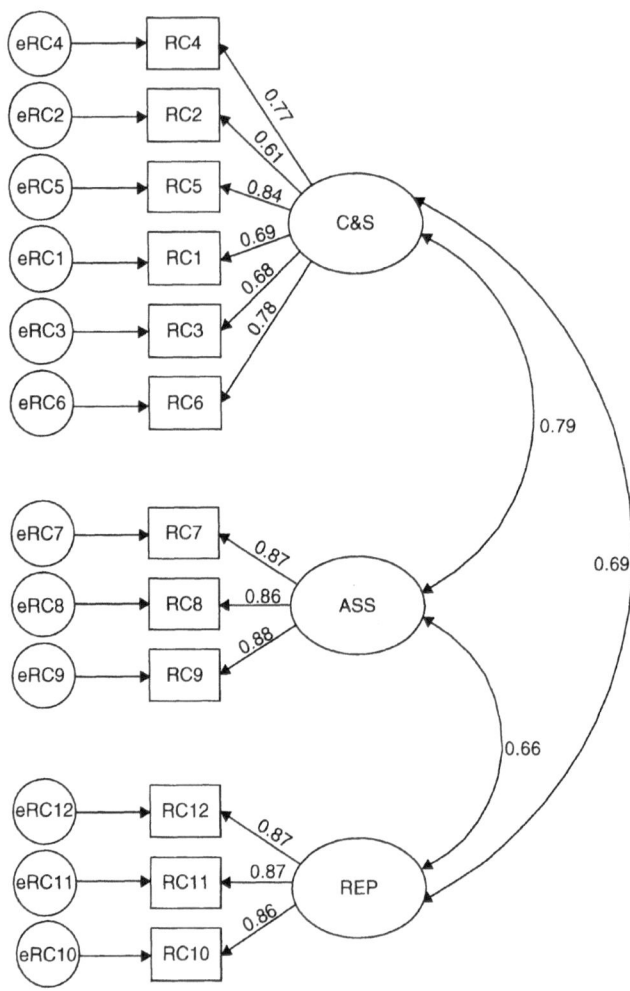

Figure 6.7 First order confirmatory factor analysis model of relational capital

value of the normal chi-square (or taking its relation to the liberty grades shown by the model) presents a value of 2.655, remaining within the advisable levels that this index required (in the range between 1 and 3).

Table 6.13 Relational capital first order model fit

Overall fit index
RMSEA
0.081

Comparative fit indexes			
GFI	AGFI	IFI	CFI
0.914	0.869	0.958	0.957

Parsimony fit index
Normal chi-square
2.655

Table 6.14 Relational capital first order regression weights (not standardized)

			Estimate	SE	RC	p value
RC6	←	C&S	1.0000			
RC3	←	C&S	0.8874	0.0808	10.9830	***
RC1	←	C&S	0.8532	0.0769	11.0946	***
RC5	←	C&S	0.9910	0.0708	14.0007	***
RC2	←	C&S	0.9474	0.0972	9.7472	***
RC4	←	C&S	1.0366	0.0823	12.5913	***
RC9	←	Ass	1.0000			
RC8	←	Ass	0.9531	0.0540	17.6435	***
RC7	←	Ass	1.0149	0.0565	17.9709	***
RC10	←	Rep	1.0000			
RC11	←	Rep	1.1280	0.0655	17.2125	***

Taking into account the unstandardized regression weights (see Table 6.14), it reflects that critical ratios (CR) are greater than 2 (the value in which the standard error is greater than zero) and that they are significant; that is, that the regression weight of every variable with regard to its factor is significantly different at a level of 0.001.

As we can see, in the model, all the regression coefficients of the measures are statistically different from zero and the *p* value of the chi-square is greater than 0.05, so the null hypothesis, which states that the estimated correlations matrix differs from the observed one only because of sampling error, cannot be rejected.

In the case of the second order CFA, again an analysis was carried out (see Figure 6.8) to check whether the dimensions proposed for relational capital fit this construct and not anothere. In the same way as for the

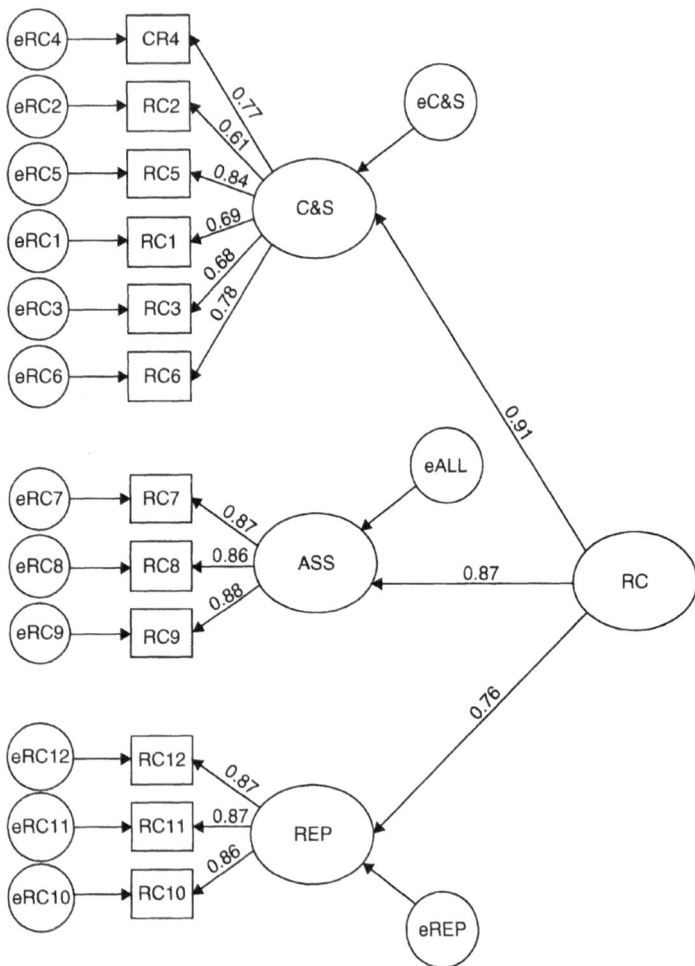

Figure 6.8 Relational capital second order confirmatory factor analysis

first order, the indicators were analysed to assess the model fit, which has the same levels as the first order relational capital model.

As for the weight of regression, Table 6.15 shows that they are significant, and that they have a few critical ratios greater than 2, as we also found in the model of the first order. So the model's fit is appropriate

Table 6.15 Relational capital second order regression weights (not standardized)

			Estimate	SE	CR	p value
C&S	←	RC	1.1838	0.1258	9.4131	***
Ass	←	RC	1.5486	0.1532	10.1052	***
Rep	←	RC	1.0000			

and it can be asserted that the relational capital dimensions are part of this kind of capital.

In the same way as for previous types of intellectual capital, it is necessary to determine the validity of the measurement model.

- *Convergent validity.* Those coefficients that are statistically significant to a confidence level of 95 per cent (t values higher than 1.96) and higher than 0.5 will show convergent validity. As we can see from previous tables, the model we have used fulfils this condition.
- *Discriminant validity.* Because it is not possible to corroborate this kind of validity with objective measures related or unrelated to the proposed model, we decided to carry out an alternative test, developed by Anderson and Gerbing (1988). This test takes the confidence interval of the correlation between each pair of critical dimensions and checks that it does not contain the value 1. If this value is not present in the intervals under discussion, it would be demonstrated that the correlation between the dimensions is far from 1, so it can be said that the pair of dimensions we have taken represent different latent concepts.

We can define reliability as the situation in which the reality we are trying to measure is indeed measured in a consistent way. In our research, reliability or the absence of random measurement errors for the latent variables has been tested by analysing the internal consistency of the items used for their definition. In this case, they have been calculated by using a compound coefficient of reliability, an alternative to the Cronbach's alpha coefficient. It is more appropriate in this case because it takes into account the different weights of each variable, without depending on the number of attributes considered for each dimension (Vandenbosch, 1996). These values are shown in Figure 6.8. As can be seen from the figure, both dimensions are greater than the minimum required level of 0.8 for exploratory studies.

Causal analysis: intellectual capital and technological innovation

Before the interpretation of the results, the normality of the dependent variables was verified, since, in this way, the suppositions that should be fulfilled by the above-mentioned linear regression residues, are assumed. Furthermore, the structure and distribution of the independent variables do not have any effect on the above-mentioned suppositions. In this sense, is not necessary to study their normality. In order to test the normality of the dependent variables, Q–Q graphs were produced, obtaining satisfactory outputs (see Appendix 4 at the end of the book).

Once the statistical tests on the variables used in the study were complete, we analysed the causal relations between intellectual capital and technological innovation, in order to test empirically the hypotheses put forward in earlier chapters.

We shall comment on the results of the linear regression on both types of technological innovation considered in the study – product and process–, using factors obtained in the EFA and CFA as independent variables and including as control variables the firms' ages and sizes.

In the first set of regressions considering product innovation, Durbin–Watson's value indicates that the residues are independent and the statistical F value shows a significant linear relation between product innovation and the set of independent variables considered in each of the elements of intellectual capital, showing the statistical validity of the proposed model.

Then the results obtained in the causal model of product innovation were analysed (see Table 6.16). As the table shows, all dimensions of the blocks of intellectual capital – human, structural and relational – have a positive effect and statistical significance on product innovation, human capital aspects being the dimension with a higher explained variance, with a R^2 of 35.1 per cent.

In addition, within human capital, all dimensions have positive and statistically significant effects on product innovation, with 'education & training' being the major one. In the case of structural capital, all dimensions have positive and statistically significant effects on product innovation, with 'culture and management commitment' being the major one. In this case, the firm's age, as a control variable, has a positive effect and statistical significance for product innovation. Finally, in the third regression, again, all relational capital dimensions have a positive effect and a statistical significance for product innovation. In this case, 'customer and supplier relationships' is the major power.

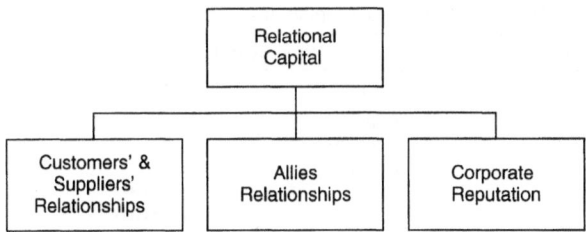

Figure 6.9 Observed model for relational capital

Table 6.16 The role of intellectual capital in product innovation

IC variables	Product innovation		
	HC	SC	RC
E&T	0.506***		
Mot	0.254***		
E&A	0.203***		
CUCO		0.426***	
R&D		0.326***	
CIT		0.241***	
C&S			0.383***
ASS			0.283***
REP			0.214***
Age	0.047	0.093*	0.059
SizeLog	−0.062	−0.063	−0.043
Model résumé			
R	0.603	0.588	0.525
R^2	0.364	0.345	0.276
Corrected R^2	0.351	0.337	0.261
T. error	0.8056	0.8140	0.8597
Durbin–Watson	1.936	1.944	2.027
F	28.033***	43.435***	18.651***

Note: Significance levels ***$p < 0.01$; **$p < 0.05$; *$p < 0.10$.

In the second set of regressions considering process innovation, again Durbin-Watson's value indicates that the residues are independent and the statistical F value shows a linear significant relation between product innovation and the set of independent variables considered in each of the elements of intellectual capital, showing the statistical validity of the proposed model.

The results obtained in the causal model of process innovation were then analysed (see Table 6.17). Despite all the dimensions of the blocks

Table 6.17   The role of intellectual capital in process innovation

IC variables	Process innovation		
	HC	SC	RC
E&T	0.389***		
Mot	0.235***		
E&A	0.194***		
CUCO		0.389***	
R&D		0.280***	
CIT		0.281***	
C&S			0.326***
ASS			0.301***
REP			0.235***
Age	0.032	0.042	0.033
SizeLog	−0.012	0.006	0.008
Model résumé			
R	0.495	0.555	0.504
R^2	0.246	0.308	0.254
Corrected R^2	0.230	0.300	0.239
T. error	0.8774	0.8366	0.8722
Durbin–Watson	1.973	1.891	1.879
F	15.945***	36.713***	16.724***

Note: Significance level ***$p < 0.01$; **$p < 0.05$; *$p < 0.10$.

of intellectual capital – human, structural and relational – having a positive effect and statistical significance for process innovation, the models have only a lower explained variance, as shown in the minor percentage of variance of the dependent variable, with structural capital aspects having the higher explained variance, with a R^2 of 30 per cent.

7
Conclusions, Limitations and Future Research

In general terms, the strategic importance that intellectual capital has in order to carry out a technological innovation has been shown from an internal and microeconomic point of view – the firm. The significance of intangible resources and capabilities– that is, knowledge – is a result of the changing environment around firms. It is important, when seeking business success, to take into account the factors that such changes have an impact upon. Specifically, the most relevant conclusions, theoretical as well as empirical, are presented below, along with the main limitations of the study and the future research directions that the study has posed.

Conclusions

Theoretical conclusions

Three main theoretical approaches have been identified in order to analyse our research phenomena, establishing the research bases. These theoretical approaches are: the resource-based view, the knowledge-based view, and the intellectual capital-based view, which make up the theoretical background for this study. Thus the specific, internal and intangible factors of firms have been studied, and their influence on technological innovation outcomes has been analysed.

Within this theoretical background, our research was framed according to two main concepts of the study that were analysed more deeply in the Chapters 2 and 3.

An exhaustive literature review of other studies on intellectual capital were carried out, bearing in mind the aim of identifying its components, and the dimensions within each of these components, in an attempt to homogenise all the ideas considered within such studies to achieve a more complete and precise model.

Because of the confusion detected in the elements of intellectual capital (Dean and Kretschmer, 2007; Alama Salazar, 2008), one of the theoretical conclusions refers to the inclusion of three main components of intellectual capital, following the most consistent classification according to the literature review. Each of these three groups, namely human capital, structural capital and relational capital, represents different types of knowledge within a firm. This structure allows a more exhaustive analysis, and a clearer identification of the sources of the capability to innovate. The inclusion of the last group of intangible assets follows the recommendations made by Acedo *et al.* (2006), showing the relational or network-based trend or current works within the resource-based view – in which that line of reasoning could be considered to be within the nature of intellectual capital – as one of the most promising theoretical perspectives for future development.

From the literature review of models of intellectual capital, some dimensions included in each of the components of intellectual capital have been identified (see Figure 7.1). This model has been completed with references not included directly in the intellectual capital field, but which provide important ideas regarding intangible firm factors. It is interesting to remark here on the importance of this reviewing effort, since

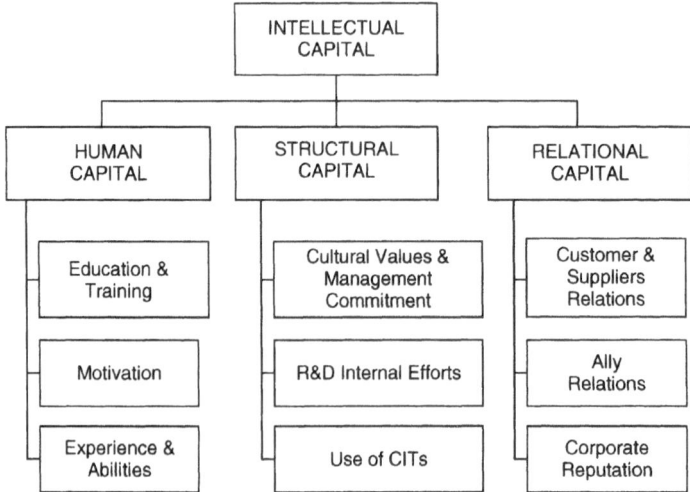

Figure 7.1 Measurement model of intellectual capital

there are very few works that have attempted to reach any conclusions on this issue (see, for example, Martínez-Torres, 2006; Alama Salazar, 2008). However, it is necessary to emphasize that our research was focused on those dimensions that provide greater interest for our main purpose: the relationship between intellectual capital and technological innovation. Thus, while there may be other possible dimensions, these have not been taken into account in order to avoid feasibility problems.

Also from this review regarding both intellectual capital literature and literature referring to intangible factors, different items have been identified, which make up each of the dimensions shown earlier. Such work required a good deal of effort to review, because studies around this topic have a clear lack of homogeneity.

A second literature review was carried out to consider technological innovation research. This wide literature remarks on the importance of ongoing adaptation to environmental changes. This is at the heart of the present competitive paradigm, in which competitive advantages are gained and sustained through subsequent innovations developed by each firm.

Through our research, a wide range of innovation typologies has been identified from studies focused on the innovation process, providing different classification criteria (see Chapter 3). In this sense, our research has focused on the typologies most accepted by the academic literature: product and process innovation. In addition, several different items have been identified in order to capture each of these innovation types.

The next step, as shown in Chapter 4, was to show arguments that referred to the causal relationships between the components of intellectual capital and the two types of technological innovation, and hypotheses were produced.

Empirical conclusions

As was mentioned earlier, different aspects of each of the studied concepts were introduced in Chapter 5, which involved a wide review of the literature regarding research related to intangible aspects, intellectual capital and technological innovation, in order to obtain a suitable measurement of the theoretical concepts developed in previous chapters.

This task was complex, because of the scarcity of empirical studies about intellectual capital, and which do not reflect clear conclusions about which items to use. It was thus necessary to analyse previous empirical literature not considered to be within the intellectual capital stream, with the aim of including an extended number of items in order to measure the constructs analysed in our study. Using this method, several knowledge factors have been measured, after carrying out a division within each of the elements of intellectual capital. Also, from some theoretical references, different dimensions that make up each of the components of intellectual capital have been identified from an empirical point of view.

The first main empirical contribution of this work – a measurement model of intellectual capital – was obtained through exploratory and confirmatory factorial analysis, reaching the following results (see Figure 7.1):

- For human capital, three dimensions were obtained: education and training; experience and abilities; and motivation.
- For structural capital, the obtained dimensions were cultural values & management commitment towards innovation; R&D internal efforts; and use of CITs.
- For relational capital, customer and supplier relationships, relationships with allies; and corporate reputation.

As the second main contribution of this book, causal relations between intellectual capital assets and technological innovation were developed. The most remarkable results are those that are derived from the comparative analysis between process and product innovation.

Therefore, it can be argued that this research has provided added value to the extant scientific literature regarding the influence of the elements of intellectual capital on a success criterion, taking technological innovation as this success-dependent variable. This work has also distinguished the elements of intellectual capital and their dimensions in order to consider different results on two types of technological innovation.

When considering each of the components of intellectual capital, results show that human capital, structural capital and relational capital have more explanatory power in the case of product innovation than process innovation. One possible explanation for this may be that, in the case of process innovation, it may involve more directly other variables that are outside our empirical research, such as firm dimension, organizational structure, or the development of total quality programmes.

To look at this in more detail, among the elements of intellectual capital that have a greater influence on product innovation, human capital leads the way with 35.1 per cent, followed by structural capital (33.7 per cent) and, finally, relational capital (26.1 per cent). Within human capital, the dimension devoted to employee 'education and training' is particularly worthwhile; within structural capital, 'culture and CEO commitment to carry out innovations' is the most remarkable; and within relational capital, vertical relationships with customers and suppliers are the most important factor for product innovation.

While it seems that the characteristics related to education and training of the human resources are key to carrying out product innovations, it is also necessary for the firm to encourage employees with a proactive attitude to accomplishment the aim of innovation, maintaining a fluent relationship with both customers and suppliers to acquire the critical knowledge that eases the innovation process.

Regarding the most important factor for product innovation according to our results – that is, the dimension 'education and training' – Hegde and Shapira (2007) argue that the continuous improvement in employees' abilities (referred to as the training of workers) is essential to maintain a high-quality workforce in order to adapt to changes in market demand and to assure the effective implementation and introduction of innovative products.

When analysing the results obtained for process innovation, structural capital (30 per cent) has a higher explanatory power than human capital (23 per cent). To look at this in detail, the most relevant dimension within structural capital is corporate culture and management commitment towards innovation, followed by R&D internal efforts, and the use of CITs. According to these results, it can be argued that cultural values, as well as the support of top management, and jointly with the effort spent on R&D activities become critical to achieve process innovations.

Perhaps one possible explanation about the relevance of structural capital – which in this proposal includes technological and organizational elements – in explaining process innovation comes from: (i) the fact that this kind of capital includes the knowledge related to productive processes, which by definition is essential for process innovation; and (ii) introducing new productive processes and/or technical changes among existing processes requires a large amount of research and development, as Huergo (2006) points out for the case of technological knowledge.

From the control variables considered in the empirical work, only firm age is significant in the model of product innovation (in general, the kinds of innovations more influenced by the components of intellectual capital), but only when structural capital is analysed. These results seem to be quite consistent with the nature of elements such as organizational culture and managerial procedures, or personal and informal relationships among employees, because they need a long enough period over which to fully develop. Work and social routines play an important role in this case, because of the nature of the involved capabilities.

Executives of high and medium-high technology firms can obtain some interesting guidelines from the above results and conclusions, considering which intangible resources and capabilities might be the most useful to obtain each type of innovation in these kinds of industries.

Limitations of the study

In this section, the main limitations of our research are shown. The first of these comes from the use of a Likert scale questionnaire, which involves the risk of subjective answers from respondents. However, this method has been applied in many studies to measure intangible resources because ad hoc questions are more suitable for collecting more detailed information on specific and internal phenomena, unlike proxies obtained from databases (Rouse and Daellenbach, 1999).

Second, two limitations come from taking only one respondent from each firm (the CEO in our case). The first limitation is that the existence of a common variance bias arises when answers about dependent and independent variables come from the same source. The second limitation makes reference to the capability of the CEO to answer the questionnaire.

Third, and regarding the measurement of the variables included of the study, three limitations can be argued: (i) instead of using the expression 'to a greater extent than my competitors' it would be more useful to include a quantitative reference point; (ii) only a single item is used for each of the identified indicators, instead of using different questions within each of them; and (iii) there may be time lags regarding the model that are not taken into account, since answers assess the intellectual capital stock and achieved innovations at the same time.

Fourth, it is necessary to emphasize that the findings of this empirical research cannot be generalized for all kinds of industry, since our sample refers only to high and medium-high technology manufacturing sectors. For this reason, implications for managerial practice can only be obtained for these industries. However, our population and sample were framed according to the recommendations of Rouse and Daellenbach (1999), because according to the industrial setting and activities, knowledge needs and usage can be different (Roos and Roos, 1997; Rouse and Daellenbach, 1999).

In addition, neither the complementarity among the different components of intellectual capital nor among the different dimensions of each element has been studied. In this sense, it would be interesting to analyse specific combinations of intellectual capital elements and dimensions; that is, to examine how they can influence each other, and the possible links among them when pursuing each of the kinds of technological innovation considered.

Finally, neither the dynamism of factors that affect firms has been taken into account, nor how firms change over time. This is a consequence of using a cross-sectional survey instead of carrying out a longitudinal study.

Future research

Because the contributions from this research can only have a limited character for the knowledge management field, it is relevant to point out some future research directions that can be followed from this point, in order to gain a deeper understanding of each of the studied aspects.

First, it would be interesting to work with objective secondary data, combining them with the data obtained in this study, as Penrose (1959) suggests. This combination is necessary because secondary sources do not provide enough information about the valuable and unique competences of a firm. For example, these data would be collected from the SABI database in order to develop objective dependent variables.

Second, a comparison between high and medium-high technology manufacturing firms can be used to understand which types of knowledge (that is, which elements of intellectual capital) are more valuable within each kind of industry, and which are not so important.

A third interesting line of research will be to analyse the phenomenon of complementarity among the different components of intellectual capital, because intellectual capital is an issue of relationships (Edvinsson, 1997), and it can therefore not be understood without considering possible links among the elements and among the dimensions within each element. In this sense, a deeper study would be worthwhile considering which elements are linked to one another, and whether that connection creates value for a firm from a technological innovation process viewpoint. In addition, this issue is emphasized by Cabrita and Bontis (2008), when they take into account, within the definition of intellectual capital, the complementarity and interactions among different types of knowledge that exist in a firm.

Finally, if possible, obtaining the same dataset over several years from the firms that compose the sample of this study would overcome the limitations of a cross-sectional study. In this way, this work would be able to add dynamic ideas to the concept of intellectual capital, as well as exploring the nature of those competitive advantages based on the innovation process.

References

Aaker, D. (1989) 'Managing Assets and Skills: The Key to a Sustainable Competitive Advantage', *California Management Review*, 31, 91–106.

Acedo, F. J., Barroso, C. and Galán, J. L. (2006) 'The Resource-Based Theory: Dissemination and Main Trends', *Strategic Management Journal*, 27, 621–36.

Achilladelis, B. and Antonakis, N. (2001) 'The Dynamics of Technological Innovation: The Case of the Pharmaceutical Industry', *Research Policy*, 30, 535–88.

Adamides, E. D. and Karacapilidis, N. (2006) 'Information Technology Support for the Knowledge and Social Process of Innovation Management', *Technovation*, 26, 50–9.

Adams, R., Bessant, J. and Phelps, R. (2006) 'Innovation Management Measurement: A Review', *International Journal of Management Reviews*, 8, 21–47.

Adner, R. (2002) 'When Are Technologies Disruptive? A Demand-based View of the Emergence of Competition', *Strategic Management Journal*, 23, 667–88.

Aiman-Smith, L., Goodrich, N., Roberts, D. and Scinta, J. (2005) 'Assessing Your Organization's Potential for Value Innovation', *Research Technology Management*, 48, 37–42.

Akgün, A. E., Keskin, H., Byrne, J. C. and Aren, S. (2007) 'Emotional and Learning Capability and Their Impact on Product Innovativeness and Firm Performance', *Technovation*, 27, 501–13.

Alama Salazar, E. M. (2008) 'Capital intelectual y Resultados Empresariales en las Empresas de Servicios Profesionales de España', Doctoral dissertation, Universidad Complutense de Madrid.

Alegre, J. and Lapiedra, R. (2005) 'Gestión del Conocimiento y Desempeño Innovador: Un Estudio del Papel Mediador del Repertorio de Competencias Distintivas', *Cuadernos de Economía y Dirección de la Empresa*, 23, 117–38.

Alegre, J. and Chiva, R. (2007) *Organizational Learning Capability, Innovation and Firm Performance* (Seville: XVII Congreso Nacional de la Asociación Científica de Economía y Dirección de la Empresa – ACEDE).

Alegre, J. and Chiva, R. (2008) 'Assessing the Impact of Organizational Learning Capability on Product Innovation Performance: An Empirical Test', *Technovation*, 28, 315–26.

Alegre, J., Lapiedra, R. and Chiva, R. (2005) *Propuesta y Validación de una Escala de Medida del Desempeño Innovador de la Empresa* (La Laguna, Tenerife, Canary Islands: XV Congreso Nacional de la Asociación Científica de Economía y Dirección de la Empresa – ACEDE).

Alegre-Vidal, J., Lapiedra-Alcamí, R. and Chiva-Gómez, R. (2004) 'Linking Operations Strategy and Product Innovation: An Empirical Study of Spanish Ceramic Tile Producers', *Research Policy*, 33, 829–39.

Almeida, P. and Phene, A. (2004) 'Subsidiaries and Knowledge Creation: The Influence of the MNC and Host Country on Innovation', *Strategic Management Journal*, 25, 847–64.

Amabile, T. M. (1998) 'How to Kill Creativity', *Harvard Business Review*, 76, 76–87.

Amabile, T. M., Barsade, S., Mueller, J. and Staw, B. (2007) 'La Conexión entre las Emociones y la Creatividad en el Trabajo', *Harvard Deusto Business Review*, 159, 36–44.

Amit, R. and Schoemaker, P. J. H. (1993) 'Strategic Assets and Organizational Rent', *Strategic Management Journal*, 14, 33–46.

Anderson, J. C. and Gerbing, D. W. (1988) 'Structural Equation Modelling in Practice: A Review and Recommended Two-step Approach', *Psychological Bulletin*, 103, 401–23.

Anderson, P. and Tushman, M. L. (1990) 'Technological Discontinuities and Dominant Designs: A Cyclical Model of Technological Change', *Administrative Science Quarterly*, 35, 604–33.

Argyris, C. (1991) 'Teaching Smart People How to Learn', *Harvard Business Review*, 69, 99–109.

Baldridge, J. V. and Burnham, R. A. (1975) 'Organizational Innovation – Individual, Organizational and Environmental Impacts', *Administrative Science Quarterly*, 20, 165–76.

Barañano, A. M., Bommer, M. and Jalajas, D. S. (2005) 'Sources of Innovation for High-tech SMEs: A Comparison of USA, Canada and Portugal', *International Journal of Technology Management*, 30, 205–19.

Barney, J. B. (1986) 'Strategic Factor Markets: Expectations, Luck, and Business Strategy', *Management Science*, 32, 1231–41.

Barney, J. B. (1991) 'Firm Resources and Sustained Competitive Advantage', *Journal of Management*, 17, 99–120.

Barney, J. B. (2001) 'Is the Resource-Based View a Useful Perspective for Strategic Management Research? Yes', *Academy of Management Review*, 26, 41–56.

Batjargal, B. (2007) 'Internet Entrepreneurship: Social Capital, Human Capital, and Performance of Internet Ventures in China', *Research Policy*, 36, 605–18.

Benito-Torres, J. L. and Varela-González, J. A. (2002) 'Influencia del Tipo de Proceso y del Grado de Novedad sobre las Actividades Ejecutadas durante el Desarrollo de Nuevos Productos', *Revista Europea de Dirección y Economía de la Empresa*, 11, 173–87.

Bierly, P. and Chakrabarti, A. (1996) 'Generic Knowledge Strategies in the U.S. Pharmaceutical Industry', *Strategic Management Journal*, 17, 123–35.

Birkinshaw, J., Hamel, G. and Mol, M. (2008) 'Management Innovation', *Academy of Management Journal*, 33, 825–45.

Blumentritt, T. and Danis, W. M. (2006) 'Business Strategy Types and Innovative Practices', *Journal of Managerial Issues*, 18, 274–91.

Boer, H. and During, W. E. (2001) 'Innovation, What Innovation? A Comparison between Product, Process and Organizational Innovation', *International Journal of Technology Management*, 22, 83–107.

Boerner, C. S., Macher, J. T. and Teece, D. J. (2001) 'Organizational Learning in Economics', in M. Dierkes, A. Berthoin-Antal, J. Child and I. Nonaka (eds), *Handbook of Organizational Learning and Knowledge* (New York: Oxford University Press), 89–117.

Bontis, N. (1996) 'There's a Price on Your Head: Managing Intellectual Capital Strategically', *Business Quarterly*, 60, 41–7.

Bontis, N. (1998) 'Intellectual Capital: An Exploratory Study that Develops Measures and Models', *Management Decision*, 36, 63–76.

Bontis, N. (1999) 'Managing Organizational Knowledge by Diagnosing Intellectual Capital: Framing and Advancing the State of the Field', *International Journal of Technology Management*, 18, 433–62.

Bontis, N. (2001) 'Assessing Knowledge Assets: A Review of the Models Used to Measure Intellectual Capital', *International Journal of Management Reviews*, 3, 41–60.

Bontis, N., Keow, W. and Richardson, S. (2000) 'Intellectual Capital and Business Performance in Malaysian Industries', *Journal of Intellectual Capital*, 1, 1–9.

Bontis, N., Crossan, M. M. and Hulland, J. (2002) 'Managing an Organizational Learning System by Aligning Stocks and Flows', *Journal of Management Studies*, 39, 437–69.

Booz, E., Allen, J. and Hamilton, C. (1982) *New Product Development for the 1980s* (New York: Booz, Allen & Hamilton Consultants).

Bossink, B. A. G. (2002) 'The Development of Co-innovation Strategies: Stages and Interaction Patterns in Interfirm Innovation', *R&D Management*, 32, 311–20.

Brennan, N. and Connell, B. (2000) 'Intellectual Capital: Current Issues and Policy Implications', *Journal of Intellectual Capital*, 1, 206–40.

Brooking, A. (1996) *Intellectual Capital: Core Asset for the Third Millennium Enterprise* (London: International Thomson Business Press).

Brooking, A. (1997) 'The Management of Intellectual Capital', *Long Range Planning*, 30, 364–5.

Brown, J. and Duguid, P. (1998) 'Organizing Knowledge', *California Management Review*, 40, 90–111.

Browne, M. W. and Cudeck, R. (1989) 'Single Sample Cross-Validation Indices for Covariance Structures', *Multivariate Behavioral Research*, 24, 445–55.

Bueno, E. (1998) 'El Capital Intangible como Clave Estratégica en la Competencia Actual', *Boletín de Estudios Económicos*, 53, 207–29.

Bueno, E., Salmador, M. P. and Rodríguez, O. (2004) 'The Role of Social Capital in Today's Economy', *Journal of Intellectual Capital*, 5, 556–74.

Cabrita, M. R. and Bontis, N. (2008) 'Intellectual Capital and Business Performance in the Portuguese Banking Industry', *International Journal of Technology Management*, 43, 212–37.

Camelo, C., Martín, F., Romero, P. and Valle, R. (2000) 'Relación entre el Tipo y el Grado de Innovación y el Rendimiento de la Empresa: Un Análisis Empírico', *Economía Industrial*, 333, 149–60.

Carmeli, A. and Tishler, A. (2004) 'The Relationships between Intangible Organizational Elements and Organizational Performance', *Strategic Management Journal*, 25, 1257–78.

Carson, E., Ranzijn, R., Winefield, A. and Marsden, H. (2004) 'Intellectual Capital. Mapping Employee and Work Group Attributes', *Journal of Intellectual Capital*, 5, 443–63.

Chandy, R. K. and Tellis, G. J. (1998) 'Organizing for Radical Product Innovation: The Overlooked Role of Willingness to Cannibalize', *Journal of Marketing Research*, 35, 474–487.

Chang, S., Chen, S. and Lai, J. (2008) 'The Effect of Alliance Experience and Intellectual Capital on the Value Creation of International Strategic Alliances', *Omega*, 36, 298–316.

Chang, Y. (2003) 'Benefits of Co-operation on Innovative Performance: Evidence from Integrated Circuits and Biotechnology Firms in the UK and Taiwan', *R&D Management*, 33, 425–37.

Chen, J., Zhu, Z. and Xie, H. Y. (2004) 'Measuring Intellectual Capital: A New Model and Empirical Study', *Journal of Intellectual Capital*, 5, 195–212.

Chesbrough, H. W. and Teece, D. J. (2003) 'Organizarse para Innovar: ¿Cuándo es Virtuoso lo Virtual?' *Harvard Deusto Business Review*, 112, 22–30.

Christensen, C. M. (1992) 'Exploring the Limits of the Technology S-Curve', *Production and Operations Management*, 1, 334–66.

Christensen, C. (1997) *The Innovator's Dilemma* (Boston, Mass.: Harvard Business School Press).

CIC (Centro de Investigación sobre la Sociedad del Conocimiento) (2003) *Modelo Intellectus: Medición y Gestión del Capital Intelectual* (Madrid: CIC-IADE – Research Center for the Knowledge Society).

Cohen, W. and Levinthal, D. (1990) 'Absorptive Capacity: A New Perspective on Learning and Innovation', *Administrative Science Quarterly*, 35, 128–52.

Cohendet P., Llerena P. and Marengo L. (2000) 'Is There a Pilot in the Evolutionary Firm?' in N. Foss and V. Mahnke (eds), *New Directions in Economic Strategy Research* (Oxford, UK: Oxford University Press), 95–115.

Conner, K. R. (1991) 'A Historical Comparison of Resource-Based Theory and Five Schools of Thought within Industrial Organizational Economics: Do We Have a New Theory of the Firm?' *Journal of Management*, 17, 121–54.

Conner, K. R. and Prahalad, C. K. (1996) 'A Resourced-based Theory of the Firm: Knowledge versus Opportunism', *Organization Science*, 7, 477–501.

Coombs, J. E. and Bierly, P. E. (2006) 'Measuring Technological Capability and Performance', *R&D Management*, 36, 421–38.

Cooper, R. G. (1985) 'Overall Corporate Strategies for New Product Development', *Industrial Marketing Management*, 14, 179–93.

Cordón-Pozo, E., García-Morales, V. J. and Aragón-Correa, J. A. (2006) 'Inter-Departmental Collaboration and New Product Development Success: A Study on the Collaboration between Marketing and R&D in Spanish High-technology Firms', *International Journal of Technology Management*, 35, 52–79.

Corso, M., Martín, A., Paolucci, E. and Pellegrini, L. (2001) 'Knowledge Management in Product Innovation: An Interpretative Review', *International Journal of Management Reviews*, 3, 341–52.

Cowan, R. and Foray, D. (1997) 'The Economics of Codification and the Diffusion of Knowledge', *Industrial and Corporate Change*, 6, 595–622.

Crossan, M. M., Lane, H. W. and White, R. E. (1999) 'An Organizational Learning Framework: From Intuition to Institution', *Academy of Management Review*, 24, 522–37.

Daellenbach, U. S., McCarthy, A. M. and Schoenecker, T. S. (1999) 'Commitment to Innovation: The Impact of Top Management Team Characteristics', *R&D Management*, 29, 199–208.

Damanpour, F. (1987) 'The Adoption of Technological, Administrative, and Ancillary Innovations: Impact of Organizational Factors', *Journal of Management*, 13, 675–88.

Damanpour, F. (1991) 'Organizational Innovation: A Meta-analysis of Effects of Determinants and Moderators', *Academy of Management Journal*, 34, 555–90.

Damanpour, F. and Evan, W. M. (1984) 'Organizational Innovation and Performance: The Problem of Organizational Lag', *Administrative Science Quarterly*, 29, 392–409.

Damanpour, F. and Gopalakrishnan, S. (1998) 'Theories of Organizational Structure and Innovation Adoption: The Role of Environmental Change', *Journal of Engineering and Technology Management*, 15, 1–24.

Damanpour, F. and Gopalakrishnan, S. (2001): 'The Dynamics of the Adoption of Product and Process Innovations in Organizations', *Journal of Management Studies*, 38, 45–65.

Danneels, E. (2002) 'The Dynamics of Product Innovation and Competences', *Strategic Management Journal*, 23, 1095–121.

Darroch, J. and McNaughton, R. (2002) 'Examining the Link between Knowledge Management Practices and Types of Innovation', *Journal of Intellectual Capital*, 3, 210–22.

Dean, A. and Kretschmer, M. (2007) 'Can Ideas Be Capital? Factors of Production in the Postindustrial Economy: A Review and Critique', *Academy of Management Review*, 32, 573–94.

De Carolis, D. (2003) 'Competencies and Inimitability in the Pharmaceutical Industry: An Analysis of their Relationship with Firm Performance', *Journal of Management*, 29, 27–50.

De Carolis, D. and Deeds, D. (1999) 'The Impact of Stocks and Flows of Organizational Knowledge on Firm Performance: An Empirical Evaluation of the Biotechnology Industry', *Strategic Management Journal*, 20, 953–68.

Deephouse, D. L. (2000) 'Media Reputation as a Strategic Resource: An Integration of Mass Communication and Resource-Based Theories', *Journal of Management*, 26, 1091–112.

De Saá, P. and Díaz, N. L. (2007) 'Incidencia de los Recursos Humanos de I + D Internos y Contratados en la Innovación', *Cuadernos de Economía y Dirección de la Empresa*, 33, 7–30.

Deward, R. D. and Dutton, J. E. (1986) 'The Adoption of Radical and Incremental Innovations: An Empirical Analysis', *Management Science*, 32, 1422–33.

Díaz, N. L., Aguiar, I. and De Saá, P. (2006) 'El Conocimiento Organizativo Tecnológico y la Capacidad de Innovación. Evidencia para la Empresa Industrial Española', *Cuadernos de Economía y Dirección de la Empresa*, 27, 33–60.

Díaz Díaz, N. L. and De Saá Pérez, P. (2007) *El Papel de los Recursos Humanos de I + D en la Absorción del Conocimiento Adquirido mediante Alianzas* (Seville: XVII Congreso Nacional de la Asociación Científica de Economía y Dirección de la Empresa – ACEDE).

Dierickx, I. and Cool, K. (1989) 'Asset Stock Accumulation and Sustainability of Competitive Advantage', *Management Science*, 35, 1504–13.

Dodgson, M. (1993) 'Organizational Learning: A Review of Some Literatures', *Organization Studies*, 14, 375–94.

Dodgson, M. and Hinze, S. (2000) 'Indicators Used to Measure the Innovation Process: Defects and Possible Remedies', *Research Evaluation*, 8, 101–14.

Dollinger, M. J., Golden, P. A. and Saxton, T. (1997) 'The Effect of Reputation on the Decision to Joint Venture', *Strategic Management Journal*, 18, 127–40.

Dossi, G. and Teece, D. J. (1993) 'Organizational Competencies and the Boundaries of the Firm', CCC Working paper No. 93-11 (Berkeley, Calif.: Center for Research in Management, University of California at Berkeley).

Dossi, G. and Marengo, L. (2000) 'On the Tangled Discourse between Transaction Cost Economics and Competence-Based Views of the Firm: Some Comments', in N. J. Foss, and , V. Mahnke (eds), *Competence, Governance, and Entrepreneurship* (New York: Oxford University Press), 80–92.

Dowling, M. J. and McGee, J. E. (1994) 'Technology Strategy and New Venture Development in the Telecommunications Industry', *Management Science*, 40, 1663–77.

Drucker, P. F. (1995) *Managing in a Time of Great Change* (New York: Truman Talley).

Dutta, S., Narasimhan, O. and Rajiv, S. (2005) 'Conceptualizing and Measuring Capabilities: Methodology and Empirical Application', *Strategic Management Journal*, 26, 277–85.

Dyer, J. H. and Singh, H. (1998) 'The Relational View: Cooperative Strategy and Sources of Interorganizational Competitive Advantage', *Academy of Management Review*, 23, 660–79.

Dzinkowski, R. (2000) 'The Measurement and Management of Intellectual Capital: An Introduction', *Management Accounting*, 77, 32–6.

EC (European Commission) (1995) *Green Paper on Innovation* (Brussels: European Commission).

Edvinsson, L. (1997) 'Developing Intellectual Capital at Skandia', *Long Range Planning*, 30, 266–373.

Edvinsson, L. and Sullivan, P. (1996) 'Developing a Model for Managing Intellectual Capital', *European Management Journal*, 14, 356–64.

Edvinsson, L. and Malone, M. (1997) *Intellectual Capital: Realizing Your Company's True Value by Finding Its Hidden Brainpower* (New York: HarperCollins).

Edvinsson, L., Dvir, R., Roth, N. and Pasher, E. (2004) 'Innovations: The New Unit of Analysis in the Knowledge Era', *Journal of Intellectual Capital*, 5, 40–58.

Egbetokun, A. A., Siyanbola, W. O., Sanni, M., Olamade, O. O., Adeniyi, A. A. and Irefin, I. A. (2009) 'What Drives Innovation? Inferences from an Industry-wide Survey in Nigeria', *International Journal of Technology Management*, 45, 123–40.

Eisenhardt, K. M. (1989) 'Making Fast Strategic Decisions in High-Velocity Environments', *Academy of Management Journal*, 32, 543–76.

Elenkov, D. S. and Manev, I. M. (2005) 'Top Management Leadership and Influence on Innovation: The Role of Sociocultural Context', *Journal of Management*, 31, 381–402.

Escorsa, P. and Valls, J. (1997) *Tecnología e Innovación en la Empresa. Dirección y Gestión* (Barcelona: Ediciones UPC).

Fernández Rodríguez, Z. and Suárez González, I. (1996): 'La Estrategia de la Empresa desde una Perspectiva Basada en los Recursos', *Revista Europea de Dirección y Economía de la Empresa*, 5, 73–92.

Foo, M., Sin, H. and Yiong, L. (2006) 'Effects of Team Inputs and Intrateam Processes on Perceptions of Team Viability and Member Satisfaction in Nascent Ventures', *Strategic Management Journal*, 27, 389–99.

Foss, N. J. (1994) 'The Biological Analogy and the Theory of the Firm: Marshall and Monopolistic Competition', *Journal of Economic Issues*, 28, 1115–36.

Foss, N. J. (1996) 'Knowledge-Based Approaches to the Theory of the Firm: Some Critical Comments', *Organization Science*, 7, 470–6.

Foss, N. J. (1997) The Classical Theory of Production and the Capabilities View of the Firm', *Journal of Economic Studies*, 24, 307–23.

Foss, N. J. and Knudsen, T. (2003) 'The Resource-based Tangle: Towards a Sustainable Explanation of Competitive Advantage', *Managerial and Decision Economics*, 24, 291–307.

Freeman, C. and Soete, L. (1997) *The Economics of Industrial Innovation* (Cambridge, Mass.: MIT Press).

Galbraith, J. K. (1969) 'The Consequences of Technology', *Journal of Accountancy*, 127, 44.

Galende, J. (2006) 'Analysis of Technological Innovation from Business Economics and Management', *Technovation*, 26, 300–11.

Galende, J. and De la Fuente, J. M. (2003) 'Internal Factors Determining a Firm's Innovative Behaviour', *Research Policy*, 32, 715–36.

Gallego, A. and Casanueva, C. (2007) *El Peso de la Cooperación en la Innovación de la Empresa Industrial Española* (Seville: XVII Congreso Nacional de la Asociación Científica de Economía y Dirección de la Empresa – ACEDE).

Galunic, D. C. and Rodan, S. (1998) 'Resource Recombinations in the Firm: Knowledge Structures and the Potential for Schumpeterian Innovation', *Strategic Management Journal*, 19, 1193–201.

Galunic, D. C. and Eisenhardt, K. M. (2001) 'Architectural Innovation and Modular Corporate Forms', *Academy of Management Journal*, 44, 1229–49.

García, R. and Calantone, R. (2002) 'A Critical Look at Technological Innovation Typology and Innovativeness Terminology: A Literature Review', *Journal of Product Innovation Management*, 19, 110–32.

García, F. and Navas, J. E. (2007) 'Explaining and Measuring Success in New Business: The Effect of Technological Capabilities on Firm Results', *Technovation*, 27, 30–46.

García, T. and Mulero, E. (2007) 'Medida de los Factores Claves del Éxito de la I + D: El Constructo y sus Dimensiones', *Cuadernos de Economía y Dirección de la Empresa*, 32, 15–48.

Gatignon, H., Tushman, M. L., Smith, W. and Anderson, P. (2002) 'A Structural Approach to Assessing Innovation: Construct Development of Innovation Locus, Type, and Characteristics', *Management Science*, 48, 1103–22.

Gee, S. (1981) *Technology Transfer, Innovation, and International Competitiveness* (New York: John Wiley).

Godfrey, P. C. and Hill, C. W. L. (1995) 'The Problem of Unobservables in Strategic Management Research', *Strategic Management Journal*, 6, 519–34.

Govindarajan, V. and Kopalle, P. K. (2006) 'Disruptiveness of Innovations: Measurement and an Assessment of Reliability and Validity', *Strategic Management Journal*, 27, 189–99.

Granstrand, O. (1998) 'Towards a Theory of the Technology-Based Firm', *Research Policy*, 27, 465–89.

Granstrand, O. (1999) *The Economics and Management of Intellectual Property: Towards Intellectual Capitalism* (Cheltenham, UK: Edward Elgar).

Grant, R. M. (1991) 'The Resource-Based Theory of Competitive Advantage: Implications for Strategy Formulation', *California Management Review*, 33, 114–35.

Grant, R. M. (1996) 'Toward a Knowledge-Based Theory of the Firm', *Strategic Management Journal*, 17, 109–22.

Grant, R. M. (1997) 'The Knowledge-Based View of the Firm: Implications for Management Practice', *Long Range Planning*, 30, 450–4.

Grant, R. M. (2006) *Dirección Estratégica. Conceptos, Técnicas y Aplicaciones*, 5th edn (Madrid: Civitas).

Grant, R. M. and Baden-Fuller, C. (2004) 'A Knowledge Assessing Theory of Strategic Alliances', *Journal of Management Studies*, 41, 61–84.

Greve, H. R. (2003) 'A Behavioral Theory of R&D Expenditures and Innovations: Evidence from Shipbuilding', *Academy of Management Journal*, 46, 685–702.

Gummesson, E. (1989) 'Nine Lessons on Service Quality', *TQM Magazine*, 1, 83–7.

Guthrie, J., Johanson, U., Bukh, P. N. and Sánchez, P. (2003) 'Intangibles and the Transparent Enterprise: New Strands of Knowledge', *Journal of Intellectual Capital*, 4, 429–40.

Guthrie, J., Petty, R., Yongvanich, K. and Ricceri, F. (2004) 'Using Content Analysis as a Research Method to Inquire into Intellectual Capital Reporting', *Journal of Intellectual Capital*, 5, 282–93.

Hagedoorn, J. and Duysters, G. (2002) 'Learning in Dynamic Inter-Firm Networks: The Efficacy of Multiple Contracts', *Organization Studies*, 23, 525–48.

Hall, R (1992) 'The Strategic Analysis of Intangible Resources', *Strategic Management Journal*, 13, 135–44.

Hall, R. (1993) 'A Framework Linking Intangible Resources and Capabilities to Sustainable Competitive Advantage', *Strategic Management Journal*, 14, 607–18.

Hamel, G. and Prahalad, C. K (1994) *Competing for the Future* (Boston, Mass.: Harvard Business School Press).

Han, J. K., Kim, N. and Srivastava, K. (1998) 'Market Orientation and Organizational Performance: Is Innovation a Missing Link?', *Journal of Marketing*, 62, 30–45.

Harris, R. G. (2001) 'The Knowledge-Based Economy: Intellectual Origins and New Economic Perspectives', *International Journal of Management Reviews*, 3, 21–40.

Hayton, J. C. (2005) 'Competing in the New Economy: The Effect of Intellectual Capital on Corporate Entrepreneurship in High-Technology New Ventures', *R&D Management*, 35, 137–55.

Hedlund, G. (1994) 'A Model of Knowledge Management and the N-Form Corporation', *Strategic Management Journal*, 15, 73–90.

Hegde, D. and Shapira, P. (2007) 'Knowledge, Technology Trajectories, and Innovation in a Developing Country Context: Evidence from a Survey of Malaysian Firms', *International Journal of Technology Management*, 40, 349–70.

Henderson, R. M. and Clark, K. B. (1990) 'Architectural Innovation: the Reconfiguration of Existing Product Technologies and the Failure of Established Firms', *Administrative Science Quarterly*, 35, 9–30.

Henderson, R. and Cockburn, I. (1994) 'Measuring Competence? Exploring Firm Effects in Pharmaceutical Research', *Strategic Management Journal*, 15, 63–84.

Hermans, R. and Kauranen. I. (2005) 'Value Creation Potential of Intellectual Capital in Biotechnology – Empirical Evidence from Finland', *R&D Management*, 35, 171–85.

Hidalgo, A. and Albors, J. (2008) 'Innovation Management Techniques and Tools: A Review from Theory and Practice', *R&D Management*, 38, 113–27.

Hill, C. W. L. and Rothaermel, F. T. (2003) 'The Performance of Incumbent Firms in the Face of Radical Technological Innovation', *Academy of Management Review*, 28, 257–74.

Hsieh, M. and Tsai, K. (2007) 'Technological Capability, Social Capital and the Launch Strategy for Innovative Products', *Industrial Marketing Management*, 36, 493–502.

Hsu, Y. H. and Fang, W. (2009) 'Intellectual Capital and New Product Development Performance: The Mediating Role of Organizational Learning Capability', *Technological Forecasting and Social Change*, 76, 664–77.

Huber, G. P. (1991) 'Organizational Learning: The Contributing Processes and the Literatures', *Organization Science*, 2, 81–115.

Huergo, E. (2006) 'The Role of Technological Management as a Source of Innovation: Evidence from Spanish Manufacturing Firms', *Research Policy*, 35, 1377–88.

Hurmelinna-Laukkanen, P., Sainio, L. and Jauhiainen, T. (2008) 'Appropiability Regime for Radical and Incremental Innovations', *R&D Management*, 38, 278–89.

Huselid, M. (1995) 'The Impact of Human Resource Management Practices on Turnover, Productivity, and Corporate Financial Performance', *Academy of Management Journal*, 38, 635–72.

I.U. Euroforum Escorial (1998) *Medición del Capital Intelectual. Modelo Intelect* (Madrid: Instituto Universitario Euroforum Escorial).

INE (2009) *Indicadores de Alta Tecnología* (Instituto Nacional de Estadística – Spanish National Statistics Institute).

Itami, H., and Roehl, T. (1987) *Mobilizing Invisible Assets* (Cambridge, Mass.: Harvard University Press).

James, E. H. (2000) 'Race-Related Differences in Promotions and Support: Underlying Effects of Human and Social Capital', *Organization Science*, 11, 493–508.

Jensen, M. B., Johnson, B., Lorenz, E. and Lundvall, B. A. (2007) 'Forms of Knowledge and Modes of Innovation', *Research Policy*, 36, 680–93.

Jiménez, D. and Sanz, R. (2006) 'Innovación, Aprendizaje Organizativo y Resultados Empresariales. Un Estudio Empírico', *Cuadernos de Economía y Dirección de la Empresa*, 29, 31–55.

Johnson, L. D., Neave, E. H. and Pazderka, B. (2002) 'Knowledge, Innovation and Share Value', *International Journal of Management Reviews*, 4, 101–34.

Joia, A. (2004) 'Are Frequent Customers Always a Company's Intangible Asset? Some Findings Drawn from an Exploratory Case Study', *Journal of Intellectual Capital*, 5, 586–601.

Kaplan, R. S. and Norton, D. P. (1992) 'The Balanced Scorecard – Measures that Drive Performance', *Harvard Business Review*, 70, 71–9.

Kaplan, R. S. and Norton, D. P. (2004) 'Measuring the Strategic Readiness of Intangible Assets', *Harvard Business Review*, 82, 52–63.

King, A. W. and Zeithaml, C. P. (2003) 'Measuring Organizational Knowledge: A Conceptual and Methodological Framework', *Strategic Management Journal*, 24, 763–72.

King, D. R., Covin, J. G. and Hegarty, W. H. (2003) 'Complementary Resources and the Exploitation of Technological Innovations', *Journal of Management*, 29, 589–606.

Knight, K. E. (1967) 'A Descriptive Model of the Intra-Firm Innovation Process', *Journal of Business*, 40, 478–96.

Knudsen, M. P. (2007) 'The Relative Importance of Interfirm Relationships and Knowledge Transfer for New Product Development Success', *Journal of Product Innovation Management*, 24, 117–38.

Koberg, C. S., Detienne, D. R. and Heppard, K. A. (2003) 'An Empirical Test of Environmental, Organizational, and Process Factors Affecting Incremental and Radical Innovation', *Journal of High Technology Management Research*, 14, 21–45.

Kogut, B. (2000) 'The Network as Knowledge: Generative Rules and the Emergence of Structure', *Strategic Management Journal*, 21, 405–25.

Kogut, B. and Zander, U. (1992) 'Knowledge of the Firm, Combinative Capabilities, and the Replication of Technology', *Organization Science*, 3, 383–97.

Kogut, B. and Zander, U. (1993) 'Knowledge of the Firm and the Evolutionary Theory of the Multinational Corporation', *Journal of International Business Studies*, 24, 625–45.

Kogut, B. and Zander, U. (1996) 'What Firms Do? Coordination, Identity and Learning', *Organization Science*, 7, 502–18.

Kotabe, M., Martin, X. and Domoto, H. (2003) 'Gaining from Vertical Partnerships: Knowledge Transfer, Relationship Duration, and Supplier Performance Improvement in the U.S. and Japanese Automotive Industries', *Strategic Management Journal*, 24, 293–316.

Kyriakopoulos, K. and De Ruyter, K. (2004) 'Knowledge Stocks and Information Flows in New Product Development', *Journal of Management Studies*, 41, 1469–98.

Lam, A. (2000) 'Tacit Knowledge, Organizational Learning and Societal Institutions: An Integrated Framework', *Organization Studies*, 21, 487–513.

Laursen, K. and Salter, A. (2006) 'Open for Innovation: The Role of Openness in Explaining Innovation Performance among U.K. Manufacturing Firms', *Strategic Management Journal*, 27, 131–50.

Lehmann, A., Overton, J. McC. and Leathwick, J. R. (1999) *GRASP: Generalized Regression Analysis and Spatial Predictions*, User's manual (Hamilton, New Zealand: Landcare Research).

Leiponen, A. (2006) 'Managing Knowledge for Innovation: The Case of Business-to-Business Services', *Journal of Product Innovation Management*, 23, 238–58.

Leitner, K. (2005) 'Managing and Reporting Intangible Assets in Research Technology Organisations', *R&D Management*, 35, 125–36.

Leonard-Barton, D. (1992) 'Core Capabilities and Core Rigidities: A Paradox in Managing New Product Development', *Strategic Management Journal*, 13, 111–25.

Lepak, D. P. and Snell, S. A. (2002) 'Examining the Human Resource Architecture: The Relationships among Human Capital, Employment, and Human Resource Configurations', *Journal of Management*, 28, 517–43.

Levinthal, D. A. and March, J. G. (1993) 'The Myopia of Learning', *Strategic Management Journal*, 14, 95–112.

Li, D., Eden, L., Hitt, M. A. and Ireland, R. D. (2008) 'Friends, Acquaintances, or Strangers? Partner Selection in R&D Alliances', *Academy of Management Journal*, 51, 315–34.

Li, H. and Atuahene-Gima, K. (2002) 'The Adoption of Agency Business Activity, Product Innovation, and Performance in Chinese Technology Ventures', *Strategic Management Journal*, 23, 469–90.

Li, T. and Calantone, R. J. (1998) 'The Impact of Market Knowledge Competence on New Product Advantage: Conceptualization and Empirical Examination', *Journal of Marketing*, 62, 13–29.

Lim, L. L. K. and Dallimore, P. (2004) 'Intellectual Capital: Management Attitudes in Service Industries', *Journal of Intellectual Capital*, 5, 181–94.

Lin, L. and Lu, I. (2007) 'Process Management and Technological Innovation: An Empirical Study of the Information and Electronic Industry in Taiwan', *International Journal of Technology Management*, 37, 178–92.

Lipparini, A. and Fratocchi, L. (1999) 'The Capabilities of the Transnational Firm: Accessing Knowledge and Leveraging Inter-firm Relationships', *European Management Journal*, 17, 655–67.

Lippman, S. A. and Rumelt R. P. (1982) 'Uncertain Imitability: An Analysis of Interfirm Differences in Efficiency under Competition.' *Bell Journal of Economics*, 13, 418–38.

Liu, P. L., Chen, W. C. and Tsai, C. H. (2005) 'An Empirical Study on the New Correlation between the Knowledge Management Method and New Product Development Strategy on Product Performance in Taiwan's Industries', *Technovation*, 25, 637–44.

Lloréns, J., Ruiz, A. and García, V. (2005) 'Influence of Support Leadership and Teamwork Cohesion on Organizational Learning, Innovation and Performance: An Empirical Examination', *Technovation*, 25, 1159–72.

Lounamaa, P. H. and March, J. G. (1987) 'Adaptive Coordination of a Learning Team', *Management Science*, 33, 107–23.

Madhok, A. (1996) 'The Organization of Economic Activity: Transaction Costs, Firm Capabilities and the Nature of Governance', *Organization Science*, 7, 577–90.

Makri, M., Lane, P. J. and Gomez-Mejia, L. R. (2006) 'CEO Incentives, Innovation, and Performance in Technology-intensive Firms: A Reconciliation of Outcome and Behavior-based Incentive Schemes', *Strategic Management Journal*, 27, 1057–80.

Mansury, M. A. and Love, J. H. (2008) 'Innovation, Productivity and Growth in US Business Services: A Firm-Level Analysis', *Technovation*, 28, 52–62.

March, J. G. (1991) 'Exploration and Exploitation in Organizational Learning', *Organization Science*, 2, 71–87.

Martín de Castro, Navas López and López Sáez (2009) 'A Resource-Based View of Corporate Reputation: A Direct Test', Working paper (Madrid: Universidad Complutense de Madrid).

Martín de Castro, G., López Sáez, P., Navas López, J. E. and Galindo Dorado, R. (2007) *Knowledge Creation Processes. Theory and Empirical Evidence from Knowledge-intensive Firms* (Basingstoke, UK: Palgrave Macmillan).

Martínez, A., Vela, M. J., Pérez, M. and De Luis, P. (2007) 'Flexibilidad e Innovación: El Efecto Moderador de la Cooperación', *Revista Europea de Dirección y Economía de la Empresa*, 16, 69–88.

Martínez-Torres, M. R. (2006) 'A Procedure to Design a Structural and Measurement Model of Intellectual Capital: An Exploratory Study', *Information & Management*, 43, 617–26.

McElroy, M. W. (2002) 'Social Innovation Capital', *Journal of Intellectual Capital*, 3, 30–9.

McEvily, S. K. and Chakravarthy, B. (2002) 'The Persistence of Knowledge-Based Advantage: An Empirical Test for Product Performance and Technological Knowledge', *Strategic Management Journal*, 23, 285–305.

McEvily, S. K., Eisenhardt, K. M. and Prescott, J. E. (2004) 'The Global Acquisition, Leverage, and Protection of Technological Competencies', *Strategic Management Journal*, 25, 713–22.

McGahan, A. M. and Porter, M. E. (1997) 'How Much Does Industry Matter, Really?' *Strategic Management Journal*, 18, 15–30.

McGahan, A. M. and Porter, M. E. (1998) 'What Do We Know about Variance in Accounting Profitability?', Working paper (Cambridge, Mass.: Harvard Business School).

McGahan, A. M. and Porter, M. E. (2002) 'What Do We Know about Variance in Accounting Profitability?', *Management Science*, 48, 834–51.

Miller, D. (1987) 'The Structural and Environmental Correlates of Business Strategy', *Strategic Management Journal*, 8, 55–76.

Miller, D. and Shamsie, J. (1996) 'The Resource-Based View of the Firm in Two Environments: The Hollywood Film Studios from 1936 to 1965', *Academy of Management Journal*, 3, 519–43.

Miller, W. L. (2006) 'Innovation Rules!', *Research Technology Management*, 49, 8–14.

Moon, Y. J. and Kym, H. G. (2006) 'A Model for the Value of Intellectual Capital', *Canadian Journal of Administrative Sciences*, 23, 25369.

Morcillo, P. (1995) *La Innovación en la Empresa: Factor de Supervivencia* (Madrid: Asociación Española de Contabilidad y Administración de Empresas – AECA).

Morcillo, P. (1997) *Dirección Estratégica de la Tecnología e Innovación. Un Enfoque de Competencias* (Madrid: Civitas).

Muñoz, A. and Cordón, E. (2002) 'Tamaño, Estructura e Innovación Organizacional', *Revista Europea de Dirección y Economía de la Empresa*, 11, 103–20.

Myers, S. and Marquis, D. G. (1969) *Successful Industrial Innovation* (Washington, DC: National Science Foundation).

Nahapiet, J. and Ghoshal, S. (1998) 'Social Capital, Intellectual Capital, and the Organizational Advantage', *Academy of Management Review*, 23, 24266.

Natti, S., Halinen, A. and Hanttu, N. (2006) 'Customer Knowledge Transfer and Key Account Management in Professional Service Organizations', *International Journal of Service Industry Management*, 17, 304–19.

Navas, J. E. (1994) *Organización de la Empresa y Nuevas Tecnologías* (Madrid: Pirámide).

Navas López, J. E. and Guerras Martín, L. A. (2002) *La Dirección Estratégica de la Empresa. Teoría y Aplicaciones*, 3rd edn (Madrid: Civitas).

Negassi, S. (2004) 'R&D Co-operation and Innovation: A Microeconometric Study on French Firms', *Research Policy*, 33, 365–84.

Nelson, R. R. (1968) 'A "Diffusion" Model of International Productivity Differences in Manufacturing', *American Economic Review*, 58, 1219–48.

Nelson, R. R. (1995) 'Recent Evolutionary Theorizing About Economic Change', *Journal of Economic Literature*, 33, 48–90.

Nelson, R. R and Winter, S. G (1982) *An Evolutionary Theory of Economic Change* (Cambridge, Mass.: Harvard University Press).

Nerkar, A. and Roberts, P. W. (2004) 'Technological and Product–Market Experience and the Success of New Product Introductions in the Pharmaceutical Industry', *Strategic Management Journal*, 25, 779–800.

Newbert, S. L. (2007) 'Empirical Research on the Resource-Based View of the Firm: An Assessment and Suggestions for Future Research", 28, 121–46.

Nieto, M. (2001) *Bases para el Estudio del Proceso de Innovación Tecnológica en la Empresa* (León: Universidad de León).

Nieto, M. and Quevedo, P. (2005) 'Absorptive Capacity, Technological Opportunity, Knowledge Spillovers, and Innovative Effort', *Technovation*, 25, 1141–57.

Nonaka, I. (1991) 'The Knowledge-Creating Company', *Harvard Business Review*, 69, 96–104.

Nonaka, I. (1994) 'A Dynamic Theory of Organizational Knowledge Creation', *Organization Science*, 5, 14–37.

Nonaka, I. and Takeuchi, H. (1995) *The Knowledge-Creating Company: How Japanese Companies Create the Dynamics of Innovation* (New York: Oxford University Press).

Noria, N. and Gulati, R. (1996) 'Is Slack Good or Bad for Innovation?', *Academy of Management Journal*, 39, 1245–64.

O'Donnell, D. and O'Regan, P. (2000) 'The Structural Dimensions of Intellectual Capital: Emerging Challenges for Management and Accounting', *Southern African Business Review*, 4, 14–20.

OECD (1992) *Frascati Manual* (Paris: Organisation for Economic Co-operation and Development).

OECD (1992) *Oslo Manual: Proposed Guidelines for Collecting and Interpreting Technological Innovation Data* (Paris: Organisation for Economic Co-operation and Development).

OECD (1997) *Oslo Manual: Proposed Guidelines for Collecting and Interpreting Technological Innovation Data*, 2nd edn (Paris: Organisation for Economic Co-operation and Development).

OECD (2006) *Oslo Manual: Guidelines for Collecting and Interpreting Innovation Data*, 3rd edn (Paris: Organisation for Economic Co-operation and Development).

Ordóñez, P. (2004) 'Measuring and Reporting Structural Capital', *Journal of Intellectual Capital*, 5, 629–47.

Patel, R. and Pavitt, K. (1995) 'Patterns of Technological Activity: Their Measurement and Interpretation', in P. Stoneman (ed.), *Handbook of the Economics of Innovation and Technological Change* (Oxford, UK: Basil Blackwell), 14–51.

Penrose, E. T. (1959) *The Theory of the Growth of the Firm* (London: Basil Backwell).

Pérez Cano, C. and Quevedo Cano, P. (2006) 'Human Resources Management and Its Impact on Innovation Performance in Companies', *International Journal of Technology Management*, 35, 11–28.

Pérez-Luño, A., Valle Cabrera, R. and Wiklund, J. (2007) 'Innovation and Imitation as Sources of Sustainable Competitive Advantage', *Management Research*, 5, 67–79.

Peteraf, M. A. (1993) 'The Cornerstones of Competitive Advantage: A Resource-Based View', *Strategic Management Journal*, 14, 179–91.

Peteraf, M. A. and Barney, J. B. (2003) 'Unravelling the Resource-Based Tangle', *Managerial and Decision Economics*, 24, 309–23.

Petty, R. and Guthrie, J. (2000) 'Intellectual Capital Literature Review. Measurement, Reporting and Management', *Journal of Intellectual Capital*, 1, 155–76.

Phene, A., Fladmoe-Lindquist, K. and Marsh, L. (2006) 'Breakthrough Innovations in the U.S. Biotechnology Industry: The Effects of Technological Space and Geographic Origin', *Strategic Management Journal*, 27, 369–88.

Pizarro, I., Real, J. C. and De la Rosa, M. D. (2007) *El Papel del Capital Humano y la Cultura Emprendedora en la Innovación* (Seville: XVII Congreso Nacional de la Asociación Científica de Economía y Dirección de la Empresa – ACEDE).

Polanyi, M. (1966) *The Tacit Dimension* (Garden City, NYk: Doubleday).

Porter, M. E. (1980) *Competitive Strategy* (New York: Free Press).

Porter, M. E. (1985) *Competitive Advantage* (New York: Free Press).

Powell, T. C. (2001) 'Competitive Advantage: Logical and Philosophical Considerations', *Strategic Management Journal*, 22, 875–88.

Powell, W. W., Koput, K. W. and Smith-Doerr, L. (1996) 'Inter-Organizational Collaboration and the Locus of Innovation: Networks of Learning in Biotechnology', *Administrative Science Quarterly*, 41, 116–45.

Prahalad, C. and Hamel, G. (1990) 'The Core Competence of the Corporation', *Harvard Business Review*, 90, 79–91.

Prajogo, D. I. and Ahmed, P. K. (2006) 'Relationships between Innovation Stimulus, Innovation Capacity, and Innovation Performance', *R&D Management*, 36, 499–515.

Priem, R. L. and Butler, J. E. (2001) 'Is the Resource-Based "View" a Useful Perspective for Strategic Management Research?', *Academy of Management Review*, 26, 22–40.

Reed, K. K., Lubatkin, M. and Srinivasan, N. (2006) 'Proposing and Testing an Intellectual Capital-based View of the Firm', *Journal of Management Studies*, 43, 867–93.

Reed, R. and DeFillippi, R. J. (1990) 'Causal Ambiguity, Barriers to Imitation, and Sustainable Competitive Advantage', *Academy of Management Review*, 15, 88–102.

Roberts, P. and Dowling, G. (2002) 'Corporate Reputation and Sustained Superior Financial Performance', *Strategic Management Journal*, 23, 1077–093.

Roos, J. (1998) 'Exploring the Concept of Intellectual Capital', *Long Range Planning*, 31, 150–3.

Roos, G. and Roos, J. (1997) 'Measuring Your Company's Intellectual Performance', *Long Range Planning*, 30, 413–26.

Rothaermel, F. T. and Hill, C. W. L. (2005) 'Technological Discontinuities and Complementary Assets: A Longitudinal Study of Industry and Firm Performance', *Organization Science*, 16, 52–70.

Rouse, M. and Daellenbach, U. (1999) 'Rethinking Research Methods for the Resource-based Perspective: Isolating Sources of Sustainable Competitive Advantage', *Strategic Management Journal*, 20, 487–94.

Rouse, M. J. and Daellenbach, U. S. (2002) 'More Thinking on Research Methods for the Resource-Based Perspective', *Strategic Management Journal*, 23, 963–7.

Rowe, L. and Boise, W. B. (1974) 'Organizational Innovation: Current Research and Evolving Concepts', *Public Administration Review*, 34, 284–93.

Rumelt, R. (1991) 'How Much Does Industry Matter?' *Strategic Management Journal*, 12, 167–85.

Russell, R. D. and Russell, C. J. (1992) 'An Examination of the Effects of Organizational Norms, Organizational Structure, and Environmental Uncertainty on Entrepreneurial Strategy', *Journal of Management*, 18, 639–56.

Saint-Onge, H. (1996) 'Tacit Knowledge: The Key to the Strategic Alignment of Intellectual Capital', *Strategy & Leadership*, 24, 10–13.

Salman, N. and Saives, A. (2005) 'Indirect Networks: An Intangible Resource for Biotechnology Innovation', *R&D Management*, 35, 203–15.

Sampson, R. C. (2007) 'R&D Alliances and Firm Performance: The Impact of Technological Diversity and Alliance Organization on Innovation', *Academy of Management Journal*, 50, 364–86.

Sánchez, R. (2001) *Knowledge Management and Organizational Competence* (New York: Oxford University Press).

Saviotti, P. P. (1998) 'On the Dynamics of Appropriability of Tacit and of Codified Knowledge', *Research Policy*, 26, 843–56.

Schulz, M. (2001) 'The Uncertain Relevance of Newness: Organizational Learning and Knowledge Flows', *Academy of Management Journal*, 44, 661–81.

Schumpeter, J. A. (1912) *Teoría del Desenvolvimiento Económico*, 3rd edn (Mexico: Fondo de Cultura Económica).

Schumpeter, J. A. (1942) *Capitalism, Socialism and Democracy* (New York: Harper and Brothers).

Scott, S. G. and Bruce, R. A. (1994) 'Determinants of Innovative Behaviour: A Path Model of Individual Innovation in the Workplace', *Academy of Management Journal*, 37, 580–607.

Shaw, J. D., Duffy, M. K., Johnson, J. L. and Lockhart, D. E. (2005) 'Turnover, Social Capital Losses, and Performance', *Academy of Management Journal*, 48, 594–606.

Skaggs, B. C. and Youndt, M. (2004) 'Strategic Positioning, Human Capital, and Performance in Service Organizations: A Customer Interaction Approach', *Strategic Management Journal*, 25, 85–99.

Smith, K. G., Smith, K. A., Olian, J. D. and Sims, H. P., Jr. (1994) 'Top Management Team Demography and Process: The Role of Social Integration and Communication', *Administrative Science Quarterly*, 39, 412–38.

Snell, S. A. and Dean, J. W., Jr. (1992) 'Integrated Manufacturing and Human Resource Management: A Human Capital Perspective', *Academy of Management Journal*, 35, 467–504.

Song, M. and Thieme, J. (2009) 'The Role of Suppliers in Market Intelligence Gathering for Radical and Incremental Innovation', *Journal of Product Innovation Management*, 26, 43–57.

Souitaris, V. (2002) 'Technological Trajectories as Moderators of Firm-Level Determinants of Innovation', *Research Policy*, 31, 877–98.

Sousa, E. (2006) Factores Determinantes de la Efectividad de la Transferencia de Conocimiento en los Acuerdos de Colaboración Universidad-Empresa, Doctoral Dissertation, Universidad Pablo de Olavide, Seville.

Spender, J. and Grant, R. M. (1996) 'Knowledge and the Firm: Overview', *Strategic Management Journal*, 17, 5–9.

Spender, J. C. (1996) 'Making Knowledge the Basis of a Dynamic Theory of the Firm', *Strategic Management Journal*, 17, 45–62.

Stewart, T. A. (1991) 'Brainpower', *Fortune*, 123, 44–50.

Stewart, T. A. (1997) *Intellectual Capital: The New Wealth of Organizations* (New York: Doubleday).

Stewart, T. A. (1998) *La Nueva Riqueza de las Naciones: El Capital Intelectual*, (Buenos Aires: Granica).

Stieglitz, N. and Heine, K. (2007) 'Innovations and the Role of Complementarities in a Strategic Theory of the Firm', *Strategic Management Journal*, 28, 1–15.

Stuart, T. (2000) 'Interorganizational Alliances and the Performance of Firms: A Study of Growth and Innovation Rates in a High-Technology Industry', *Strategic Management Journal*, 21, 791–811.

Subramaniam, M. and Youndt, M. A. (2005) 'The Influence of Intellectual Capital on the Types of Innovative Capabilities', *Academy of Management Journal*, 48, 450–63.

Sullivan, P. H. (2001) *Rentabilizar el Capital Intelectual. Técnicas para Optimizar el Valor de la Innovación* (Buenos Aires: Paidós).

Sveiby, K. E. (1997) *The New Organizational Wealth: Managing and Measuring Knowledge-based Assets* (San Francisco: Berrett Koehler).

Swart, J. (2006) 'Intellectual Capital: Disentangling an Enigmatic Concept', *Journal of Intellectual Capital*, 7, 136–59.

Teece, D. (1998) 'Capturing Value from Knowledge Assets: The New Economy, Markets for Know-How and Intangible Assets', *California Management Review*, 40, 55–79.

Teece, D. J. (1986) 'Profiting from Technological Innovation: Implications for Integration, Collaboration, Licensing and Public Policy', *Research Policy*, 15, 285–305.

Teece, D. J. (2000) *Managing Intellectual Capital* (Oxford, UK: Oxford University Press).

Teece, D. J. and Pisano, G. (1994) 'The Dynamic Capabilities of Firms: An Introduction', *Industrial Corporate Change*, 3, 537–56.

Teece, D. J., Pisano, G. and Shuen, A. (1997) 'Dynamic Capabilities and Strategic Management', *Strategic Management Journal*, 18, 509–33.

Thompson, V. A. (1965) 'Bureaucracy and Innovation', *Administrative Science Quarterly*, 10, 1–20.

Tidd, J. (2001) 'Innovation Management in Context: Environment, Organization and Performance', *International Journal of Management Review*, 3, 169–83.

Tippins, M. and Sohi, R. (2003) 'IT Competency and Firm Performance: Is Organizational Learning a Missing Link?', *Strategic Management Journal*, 24, 745–61.

Tödtling, F., Lehner, P. and Kaufmann, A. (2009) 'Do Different Types of Innovation Rely on Specific Kinds of Knowledge Interactions?', *Technovation*, 29, 59–71.

Tsai, W. (2001) 'Knowledge Transfer in Intraorganizational Networks: Effects of Network Position and Absorptive Capacity on Business Unit Innovation and Performance', *Academy of Management Journal*, 44, 996–1004.

Tsai, W. and Ghoshal, S. (1998) 'Social Capital and Value Creation: The Role of Intrafirm Networks', *Academy of Management Journal*, 41, 464–76.

Tsang, E. W. K. (2002) 'Acquiring Knowledge by Foreign Partners for International Joint Ventures in a Transition Economy: Learning-by-Doing and Learning Myopia', *Strategic Management Journal*, 23, 835–54.

Tseng, C. and Goo, Y. J. (2005) 'Intellectual Capital and Corporate Value in an Emerging Economy: Empirical Study of Taiwanese Manufacturers', *R&D Management*, 35, 187–201.

Tushman, M. and Nadler, D. (1986) 'Organizing for Innovation', *California Management Review*, 28, 74–92.

Ulrich, D. (1998) 'Intellectual Capital = Competence x Commitment', *Sloan Management Review*, 39, 15–26.

Un, C.A. and Cuervo-Cazurra, A. (2004) 'Strategies for Knowledge Creation in Firms', *British Journal of Management*, 15, S27–S41.

Vandenbosch, M. B. (1996) 'Confirmatory Compositional Approaches to the Development of Product Spaces', *European Journal of Marketing*, 30, 23–46.

Van de Ven, A. H. (1986) 'Central Problems in the Management of Innovation', *Management Science*, 32, 590–607.

Verdú-Jover, A. J, Lorens-Montes, J. F. and García-Morales, V. J. (2005) 'Flexibility, Fit and Innovative Capacity: An Empirical Examination', *International Journal of Technological Management*, 30, 131–46.

Verganti, R. (2008) 'Design, Meanings, and Radical Innovation: A Metamodel and a Research Agenda', *Journal of Product Innovation Management*, 25, 436–56.

Vicente-Lorente, J. D. (2001) 'Specificity and Opacity as Resource-Based Determinants of Capital Structure: Evidence for Spanish Manufacturing Firms', *Strategic Management Journal*, 22, 157–77.

Wang, C. L. and Ahmed, P. K. (2004) 'The Development and Validation of the Organisational Innovativeness Construct Using Confirmatory Factor Analysis', *European Journal of Innovation Management*, 7, 303–13.

Wang, H., Yen, Y., Tsai, C. and, Lin, Y. (2008) 'An Empirical Research on the Relationships between Human Capital and Innovative Capability: A Study on Taiwan's Commercial Banks', *Total Quality Management and Business Excellence*, 19, 1189–205.

Wang, W. and Chang, C. (2005) 'Intellectual Capital and Performance in Causal Models: Evidence from the Information Technology Industry in Taiwan', *Journal of Intellectual Capital*, 6, 222–36.

Wernerfelt, B. (1984) 'A Resource-based View of the Firm', *Strategic Management Journal*, 5, 171–80.

Williams, A. (1999) *Creativity, Invention and Innovation* (Sydney: Allen & Unwin).

Wu, S., Lin, L. and Hsu, M. (2007) 'Intellectual Capital, Dynamic Capabilities and Innovative Performance of Organizations', *International Journal of Technology Management*, 39, 279–96.

Wu, W., Chang, M. and Chen, C. (2008) 'Promoting Innovation through the Accumulation of Intellectual Capital, Social Capital, and Entrepreneurial Orientation', *R&D Management*, 38, 265–77.

Yam, R. C. M., Cheng, J., Fai, K. and Tang, E. P. Y. (2004) 'An Audit of Technological Innovation Capabilities in Chinese Firms: Some Empirical Findings in Beijing, China', *Research Policy*, 33, 1123–40.

Yin, R. (1993) *Applications of Case Study Research* (Beverly Hills, Calif.: Sage).

Yli-Renko, H., Autio, E. and Sapienza, H. J. (2001) 'Social Capital, Knowledge Acquisitions, and Knowledge Exploitation in Young Technology-based Firms', *Strategic Management Journal*, 22, 587–613.

Youndt, M. A., Subramaniam, M. and Snell, S. A. (2004) 'Intellectual Capital Profiles: An Examination of Investments and Returns', *Journal of Management Studies*, 41, 335–61.

Zack, M. H. (1999) 'Developing a Knowledge Strategy', *California Management Review*, 41, 125–45.

Zahra, S. A. and Covin, J. G. (1993) 'Business Strategy, Technology Policy and Firm Performance', *Strategic Management Journal*, 14, 451–78.

Zaltman, G., Duncan, R. and Holbeck, J. (1973) *Innovation and Organizations* (New York: John Wiley).

Zander, U. and Kogut, B. (1995) 'Knowledge and the Speed of the Transfer and Imitation of Organizational Capabilities', *Organization Science*, 6, 76–92.

Zárraga, C. and Bonache, J. (2005) 'The Impact of Team Atmosphere on Knowledge Outcomes in Self-managed Teams', *Organization Studies*, 26, 661–81.

Zárraga, C. and De Saá, P. (2005) 'Comunidades de Práctica: Equipos de Trabajo para la Gestión del Conocimiento', *Revista Europea de Dirección y Economía de la Empresa*, 14, 145–58.

Zheng, W. (2008) 'A Social Capital Perspective of Innovation from Individuals to Nations: Where Is Empirical Literature Directing Us?', *International Journal of Management Reviews*. Published online: 20 November. DOI: 10.1111/j.1468-2370.2008.00247.x.

Zmud, R. W. (1984) 'An Examination of "Push–Pull" Theory Applied to Process Innovation in Knowledge Work', *Management Science*, 30, 727–38.

Appendix 1: Letter Sent to Top Management and Questionnaire (in Spanish)

Estimado Sr./Sra.:

Mi nombre es Miriam Delgado Verde, soy estudiante de doctorado en la Universidad Complutense de Madrid y actualmente estoy realizando la tesis doctoral, la cual trata sobre Capital Intelectual e Innovación Tecnológica.

Le envío esta carta para solicitarle su ayuda en el estudio que estoy llevando acabo en empresas manufactureras de alta tecnología españolas. En este sentido, le agradecería profundamente su participación en esta investigación y apreciaría su tiempo y esfuerzo en la realización del cuestionario que le adjunto.

Dicha investigación está financiada por la Comunidad de Madrid, quedando recogida dentro del proyecto "Influencia de los activos y flujos de conocimiento en los resultados empresariales", dirigido por José Emilio Navas López, Catedrático de la Universidad Complutense de Madrid.

Además, he realizado una estancia de investigación en el Manchester Institute of Innovation Research (University of Manchester), con el objetivo de mejorar la calidad de mi proyecto.

Los resultados de mi tesis doctoral serán publicados en un libro que lleva como título "An Intellectual Capital View of Technological Innovation". Como agradecimiento a su colaboración, su nombre y el nombre de su empresa aparecerán en la lista de expertos consultados en el pretest.

Por favor, el cuestionario completado puede ser enviado a la siguiente dirección de correo:

miriamdv@ccee.ucm.es
o al fax **91 394 23 71**

Le garantizo que los datos no serán utilizados para otros fines, ni facilitados a otras empresas o personas. Además, sus datos serán procesados estadísticamente a nivel agregado.

Por favor, si tiene alguna pregunta no dude en contactar conmigo en el teléfono 91 394 29 71 o por mail. Le agradezco de antemano su colaboración, esfuerzo y tiempo.

Atentamente,
Miriam Delgado

QUESTIONNAIRE (sent in Spanish)

HUMAN CAPITAL	– Agree +
My company allocates resources (money, time, etc.) to employees training to a greater extent than my competitors	1 2 3 4 5 6 7
In my company, the percentage of people who receives training is higher than my competitors	1 2 3 4 5 6 7
In my company, the percentage of people with a superior degree (bachelor, engineer, masters, etc.) is higher than my competitors	1 2 3 4 5 6 7
In my company, the percentage of jobs filled by internal promotion is higher than my competitors	1 2 3 4 5 6 7
The experience our employees have is appropriate to carry out their work satisfactorily	1 2 3 4 5 6 7
Our employees have abilities that are widely considered to bethe best in our industry	1 2 3 4 5 6 7
Our employees develop new ideas and knowledge	1 2 3 4 5 6 7
Generally speaking, our employees are satisfied with the company	1 2 3 4 5 6 7
Employees are compromised and they maintain a high sense of responsibility with the company.	1 2 3 4 5 6 7

STRUCTURAL CAPITAL	– Agree +
My company encourages creativity, innovation and/ or the development of new ideas	1 2 3 4 5 6 7
A common system of values, beliefs and objectives exists in my company, directed towards innovation	1 2 3 4 5 6 7
My company promotes experimentation and innovation as ways to enhance processes	1 2 3 4 5 6 7
Often, managers involve employees in important decision-making processes	1 2 3 4 5 6 7
In my company, managers support and lead the innovation process	1 2 3 4 5 6 7
Managers share similar beliefs about the future management of this firm	1 2 3 4 5 6 7

STRUCTURAL CAPITAL *(Continued)*	– Agree +
Much of the firm's knowledge of processes, systems and structures is contained in databases, the intranet, electronic files, etc.	1 2 3 4 5 6 7
My company uses ICT, which allow it to to learn from past situations, thus improving employees' learning and experience	1 2 3 4 5 6 7
My company prefers to use ICT for communication, co-ordination and information diffusion	1 2 3 4 5 6 7
In my company, the average percentage of R&D employees is one of highest in the industry	1 2 3 4 5 6 7
In my company, the average of R&D costs with respect to sales is one of highest in the industry	1 2 3 4 5 6 7
My company has a formalized R&D department	1 2 3 4 5 6 7

RELATIONAL CAPITAL	– Agree +
My company obtains from customers' portfolios much valuable information on market needs and tendencies	1 2 3 4 5 6 7
Employees of my company work jointly with customers to develop solutions	1 2 3 4 5 6 7
The customer base of my company is one of the best in our industry	1 2 3 4 5 6 7
Employees of my company work jointly with suppliers in order to develop solutions	1 2 3 4 5 6 7
In recent years, my company has improved the quality and design of products and processes through relationships with our suppliers	1 2 3 4 5 6 7
The supplier base of my company is one of the best in our industry	1 2 3 4 5 6 7
Employees of my company work jointly with allies in order to develop solutions	1 2 3 4 5 6 7
In recent years, my company has improved product and process quality and design through relationships with allies	1 2 3 4 5 6 7

RELATIONAL CAPITAL (Continued)	– Agree +
The allies base of my company is one of the best in our industry	1 2 3 4 5 6 7
The reputation of my company with respect to product quality is one of the best in the industry	1 2 3 4 5 6 7
The managerial reputation of my company is one of the best in the industry	1 2 3 4 5 6 7
The financial reputation of my company is one of the best in the industry	1 2 3 4 5 6 7

TECHNOLOGICAL INNOVATION	– Agree +
In the last three years, the number of product innovations developed by my company is higher than my competitors'	1 2 3 4 5 6 7
The percentage of sales with respect to new products, against total sales, is higher than my competitors'	1 2 3 4 5 6 7
In the last three years, the number of new products with respect to my product portfolio is higher than that of my competitors	1 2 3 4 5 6 7
The number of process innovations introduced by my company is higher than that of my competitors, in the last three years.	1 2 3 4 5 6 7
The new processes introduced by my company in the last three years have led to a reduction in the manufacturing cycle and/or an improvement in productive flexibility	1 2 3 4 5 6 7
The new processes introduced by my company in the last three years have led to a reduction in production costs	1 2 3 4 5 6 7

Appendix 2: Sample

1	A. AUXILIAR CARROCERA S.A.
2	AGFA GEVAERT S.A.
3	AGUIRREGOMEZCORTA Y MENDICUTE S.A.
4	AISCAN S.L.
5	AKZO NOBEL CAR REFINISHES S.L.
6	AKZO NOBEL PACKAGING COATINGS S.A.
7	ALCALIBER S.A.
8	ALTAN INNOVACION S.L.
9	ALZA S.L.
10	APLICACIONES MECANICAS VALVULAS INDUSTRIALES S.A.
11	ARAGONESA DE VEHICULOS S.A.
12	ARAGONESAS AGRO S.A.
13	ARIES ESTRUCTURAS AEROESPACIALES S.A.
14	ASCENSORS EBYP S.A.
15	ASCENSORS EBYP S.A.
16	ASIENTOS DE CASTILLA LEON S.A.
17	ASIENTOS DEL NORTE S.A.
18	ASTILLEROS CANARIOS S.A.
19	ASTILLEROS DE MALLORCA S.A.
20	AUSA CENTER S.L.
21	BAC VALVES S.A.
22	BASELL POLIOLEFINAS IBERICA S.L.
23	BASF COATINGS S.A.
24	BERMAQ S.A.
25	BEZARES S.A.
26	BIOKIT S.A.
27	BITRON INDUSTRIE ESPANA S.A.
28	BTESA
29	CABLES RCT S.A.
30	CAMAR INDUSTRIAL S.A.
31	CANCELAS NAVAL S.L.
32	CARENAGA SOCIEDAD ANONIMA LABORAL

33	CARLOS SILVA S.A.
34	CARPINTERIA NAVAL NIETO S.L.
35	CASPLE S.A.
36	CAUCHO METAL PRODUCTOS II S.L.
37	COA CONSTRUCCIONES OCCIDENTAL DE ANDALUCIA S.A.
38	COCEMFE-TOLEDO SERVICIOS MULTIPLES S.L.
39	COGNIS IBERIA S.A.
40	COLOR ESMALT S.A.
41	COMERCIAL GODEMA S.L.
42	COMPONENTES AERONAUTICOS COASA S.A.
43	COMPONENTES MECANICOS S.A.
44	CONDUCTORES DEL CINCA S.A.
45	CONSTRUCCIONES GRAVALOS S.A.
46	CONSTRUCCIONES MECANICAS ARAGONESAS S.A.
47	CONSTRUCCIONES MECANICAS CABALLE, S.A.
48	CONSTRUCCIONES NAVALES P FREIRE S.A.
49	CONSTRUCTORA DE TRANSFORMADORES DE DISTRIBUCION COTRADIS S.L.
50	CONTADORES DE AGUA DE ZARAGOZA S.A.
51	COPO FEHRER S.A.
52	COSMETICA COSBAR S.L.
53	CRETA PRINT S.L.
54	CSC ELECTRONIK S.L.
55	DATATRONICS S.A.
56	DEMAG CRANES & COMPONENTS SOCIEDAD ANONIMA.
57	DENTAID S.L.
58	DERMOFARM S.A.
59	DIC COATINGS S.L.
60	DISTRIBUCIONES MAGANA S.L.
61	DMI COMPUTER S.A.
62	DRAKA COMTEQ IBERICA S.L.
63	DUPLICO 2000 S.L.
64	ELECTROMECANICOS VIVEIRO S.A.
65	ELECTROZEMPER S.A.

66	ELEVE S.L.
67	ELSTER IBERCONTA S.A.
68	ENCLAVAMIENTOS Y SENALIZACION FERROVIARIA ENYSE S.A.
69	ENGRANAJES GRINDEL S.A.L.
70	ESPECIAL GEAR TRANSMISSIONS SOCIEDAD ANONIMA.
71	ETRA NORTE SOCIEDAD ANONIMA.
72	EUPINCA S.A.
73	EUROMED S.A.
74	F B TECNICOS ASOCIADOS S.A.
75	F M INDAUXI S.L.
76	FABRICACION COMPONENTES MOTOCICLETAS S.A.
77	FABRICACION Y REPARACION DE BUSES S.L.
78	FABRICACION Y SERVICIOS FASE S.L.
79	FACTORIAS JULIANA S.A.
80	FACTORIAS VULCANO S.A.
81	FAGOFRI S.A.
82	FAGOR ARRASATE S.C.L.
83	FARMAPROJECTS S.A.
84	FIT AUTOMOCION S.A.
85	FLUIDOCONTROL, S.A.
86	FRAN S.A.
87	FRANCISCO CARDAMA, S.A.
88	FRENOS IRUNA S.A.L
89	FUCHS LUBRICANTES S.A.
90	FUSELAJES AERONAUTICOS S.A.
91	GERMAINE DE CAPUCCINI S.A.
92	GESTAMP VIGO S.A.
93	GESTIONES ESTUDIOS Y REALIZACIONES S.A.
94	GEYSER GASTECH S.A.
95	GIVAUDAN IBERICA S.A.
96	GRUPO ANTOLIN ARAGUSA S.A.
97	GRUPO ANTOLIN-PALENCIA S.L.
98	GRUPO EPELSA S.L.
99	HAGER SISTEMAS S.A.
100	HIAB CRANES S.L.

101	HISPAVIC IBERICA S.L.
102	HITECSA AIRE ACONDICIONADO S.L.
103	HOLZMA PLATTENAUFTEILTECHNIK S.A.
104	IBERCHEM S.A.
105	IBERICA DE APARELLAJES S.L.
106	IBERSA DE PINTURAS SOCIEDAD LIMITADA.
107	ILUSOL S.A.
108	IMI CORNELIUS ESPANA S.A.
109	INAEL ELECTRICAL SYSTEMS S.A.
110	INDALVA S.L.
111	INDUSTRIA AUXILIAR ALAVESA S.A.
112	INDUSTRIAS LAGUN ARTEA S.L.
113	INDUSTRIAS PROA, S.A.
114	INDUSTRIAS PUIGJANER S.A.
115	INDUSTRIAS ZAMARBU S.A.
116	INMESOL S.L.
117	INSONORIZANTES PELZER S.A.
118	INTERFACOM S.A.
119	INTERQUIM S.A.
120	ISOWAT MADE S.L.
121	JOCU S.L.
122	JOFEMAR S.A.
123	JOHN DEERE IBERICA S.A.
124	JOVENES INDUSTRIALES METALURGICOS COMPONENTES ARAHALENSES S.L.
125	KITZ CORPORATION OF EUROPE S.A.
126	KLUBER LUBRICATION GMBH IBERICA SOCIEDAD EN COMANDITA.
127	KOROTT S.L.
128	LABORATORIO ALDO UNION S.A.
129	LABORATORIOS COSMODENT S.L.
130	LABORATORIOS DENTALES BETICOS S.L.
131	LABORATORIOS LETI S.L.
132	LAZARO ITUARTE INTERNACIONAL S.A.
133	LOREFAR S.L.
134	LUMICAN S.A.

135	M TORRES DISENOS INDUSTRIALES S.A.
136	M TORRES OLVEGA INDUSTRIAL S.L.
137	M Z IMER S.A.
138	MAC PUAR SERVICIOS INDUSTRIALES S.L.
139	MAFLOW COMPONENTS IBERICA S.L.
140	MANUEL ALMEIDA PINTO S.L.
141	MANUFACTURAS ELECTRICAS S.A.
142	MAQUINARIA AGRICOLA EL LEON S.A.
143	MATEU Y SOLE S.A.
144	MECAFRAN S.L.
145	MECANIZADOS CARDIEL S.L.
146	MECANIZADOS INDUSTRIA AUXILIAR S.A.
147	MECANOVA S.A.
148	MEDEX SOCIEDAD LIMITADA
149	MESSER IBERICA DE GASES S.A.
150	METALURGICA COMPANIA INDUSTRIAL MIRANDESA S.L.
151	METEC MOTRIC, S.A.
152	MIGUELEZ S.L.
153	MISATI S.L.
154	MIXER & PACK S.L.
155	MONTAJES NOVARUE S.L.
156	MUEBLES MONTIEL S.L.
157	NAUTICA VARADERO MAZARRON S. L.
158	NAVAL NERVION SOCIEDAD LIMITADA.
159	NODOSA S.L.
160	NOVA RECYD S.A.
161	OPTRAL S.A.
162	ORNALUX S.A.
163	PAYPER S.A.
164	PELZER DEL NORTE S.L.
165	PEMCO ESMALTES S.L.
166	PERFECTO Y PEDRO S.A.
167	PERFUMES LOEWE S.A.
168	PINTURAS BLATEM S.L.
169	PINTURAS CIN CANARIAS S.A.

170	PINTURAS HEMPEL S.A.
171	PIROTECNIA IGUAL S.A.
172	PLASTIC 7 A SOCIEDAD LIMITADA
173	PLASTRO IRRIGATION IBERICA S.L.
174	POLISUR 2000 S.A.
175	POLYONE ESPANA S.L.
176	PRAXAIR ESPANA S.L.
177	PREDAN S.A.
178	PRIMA RAM S.A.
179	PRODUCTOS CITROSOL S.A.
180	PRODUCTOS DIEZ S.A.
181	PRODUCTOS Q P SOCIEDAD ANONIMA.
182	QUIMICA INDUSTRIAL MEDITERRANEO S.L.
183	QUIMICA SINTETICA S.A.
184	QUIMICAS ORO S.A.
185	QUIMICER S.A.
186	RECREATIVOS FRANCO S.A.
187	REDOR SOCIEDAD LIMITADA
188	REFTRANS S.A.
189	REVENGA INGENIEROS S.A.
190	RIEJU S.A.
191	RIERA NADEU S.A.
192	RIPERLAMP S.A.L.
193	ROBERT BOSCH ESPAÑA FABRICA CASTELLET S.A.
194	SAFT POWER SYSTEMS IBERICA SL.
195	SALTEC EQUIPOS PARA LA CONSTRUCCION S.A.
196	SALTO SYSTEMS S.L.
197	SANDOZ INDUSTRIAL PRODUCTS S.A.
198	SANTANA MOTOR S.A.
199	SEAT SPORT S.A.
200	SGL CARBON S.A.
201	SLUZ S.L.
202	SMC-FOSECO PRODUCTOS FUNDICION SOCIEDAD ANONIMA.
203	SOCIEDAD INDUSTRIAL DE MAQUINARIA ANDALUZA S.A.
204	SOLDADURA Y MONTAJE LUARTO S.L.

205	SOLVAY QUIMICA S.L.
206	SPEC S.A.
207	SUMINISTROS PARA FERROCARRILES Y TRANVIAS S.A.
208	TALLERES CARRAL, S.L.
209	TALLERES FORO S.A.
210	TALLERES ORDUNA SL
211	TALLERES ZITRON S.A.
212	TALLERS GILI 98 S.L.
213	TALLIN S.A.
214	TEAIS S.A.
215	TECNICAS DE REPARACIONES ESPECIALES Y SERVICIOS ASTURIAS S.A.
216	TECNICAS HIDRAULICAS S.A.
217	TECNICAS MODULARES E INDUSTRIALES S.A.
218	TECNOBIT S.L.
219	TECNOCOM TELEFONIA Y REDES S.L.
220	TECNOLOGIA ELECTRONICA DEL RIPOLLES S.L.
221	TECNOTRANS BONFIGLIOLI S.A.
222	TEDEC MEIJI FARMA S.A.
223	TERMOVEN S.L.
224	THOMIL S.A.
225	THYSSENKRUPP ELEVADORES S.L.
226	TINCASUR SUR S.L.
227	TOLL MANUFACTURING SERVICES SL.
228	TOSCANO LINEA ELECTRONICA SOCIEDAD LIMITADA.
229	TRACTEL IBERICA S.A.
230	TRADE CORPORATION INTERNATIONAL S.A.
231	TRADINSA INDUSTRIAL S.L.
232	TRELLEBORG FLUID IBERICA S.A.
233	TRETY S.A.
234	TRINEON S.L.
235	TRUMPF MAQUINARIA S.A.
236	TRW AUTOMOTIVE ESPANA S.L.
237	UBE CHEMICAL EUROPE S.A.
238	ULTRAFILTER S.L.

239	UNIVERSAL DE DESARROLLOS ELECTRONICOS S.A.
240	URIACH AQUILEA OTC S.L.
241	UTIFORM TECHNOLOGIES S.L.
242	UTILAR IBERIA S.A.
243	VALENCIA MODULOS DE PUERTA S.L.
244	VALENCIANA DE RECUBRIMIENTOS S.A.
245	VERKOL S.A.
246	VIZUETE S.L.
247	VOSSLOH ESPANA SOCIEDAD ANONIMA.
248	WAT DIRECCIONES S.A.
249	ZAMBON S.A.
250	ZORROTZ COMERCIAL S.A.
251	ZUMEX MAQUINAS Y ELEMENTOS S.A.

Appendix 3: Q–Q graphs

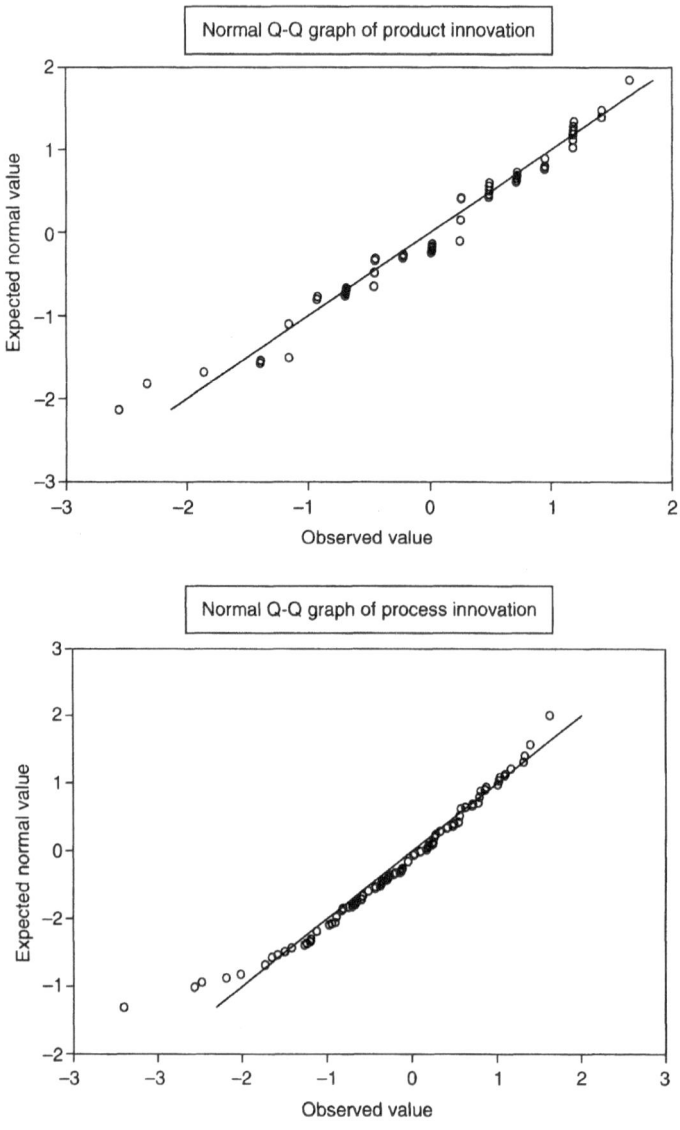

Index